This is a descriptive semantic study of taste terms in modern spoken Japanese, combining the theoretical approach of the lexical semantic field with methodological procedures developed in anthropological linguistics. Through an investigation of the range of vocabulary available for the description and evaluation of tastes in the language, Dr Backhouse presents a sensitive elucidation of the structure of central Japanese taste terms, their extended meanings, and other linguistic properties. He explores important semantic issues, such as the relation between evaluative and descriptive meaning, and the intralingual mechanisms at work in metaphor. He also draws illuminating connections between the lexical field of colour and that of taste. As the first extended treatment of the taste terms of a language other than English, Dr Backhouse's study will be welcomed not only by linguists but also by anthropologists and psychologists with an interest in this neglected area of the language of perception.

CAMBRIDGE STUDIES IN LINGUISTICS
SUPPLEMENTARY VOLUME

The Lexical Field of Taste

In this series

52 MICHAEL S. ROCHEMONT and PETER W. CULICOVER: English focus constructions and the theory of grammar
53 PHILIP CARR: Linguistic realities: an autonomist metatheory for the generative enterprise
54 EVE SWEETSER: From etymology to pragmatics: metaphorical and cultural aspects of semantic structure
55 REGINA BLASS: Relevance relations in discourse: a study with special reference to Sissala
56 ANDREW CHESTERMAN: On definiteness: a study with special reference to English and Finnish
57 ALESSANDRA GIORGI and GUISEPPE LONGOBARDI: The syntax of noun phrases: configuration, parameters and empty categories
58 MONIK CHARETTE: Conditions on phonological government
59 M. H. KLAIMAN: Grammatical voice
60 SARAH M. B. FAGAN: The syntax and semantics of middle constructions: a study with special reference to German
61 ANJUM P. SALEEMI: Universal grammar and language learnability
62 STEPHEN R. ANDERSON: A-morphous morphology
63 LESLEY STIRLING: Switch reference and discourse representation
64 HENK J. VERKUYL: A theory of aspectuality: the interaction between temporal and atemporal structure
65 EVE V. CLARK: The lexicon in acquisition
66 ANTHONY R. WARNER: English auxiliaries: structure and history
67 P. H. MATTHEWS: Grammatical theory in the United States from Bloomfield to Chomsky
68 LJILJANA PROGOVAC: Negative and positive polarity: a binding approach
69 R. M. W. DIXON: Ergativity
70 YAN HUANG: The syntax and pragmatics of anaphora: a study with special reference to Chinese
71 KNUD LAMBRECHT: Information structure and sentence form: topic, focus, and the mental representations of discourse referents

Supplementary volumes

RUDOLF P. BOTHA: Form and meaning in word formation: a study of Afrikaans reduplication
AYHAN AKSU-KOÇ: The acquisition of aspect and modality: the case of past reference in Turkish
MÍCHEÁL Ó SAIDHAIL: Modern Irish: grammatical structure and dialectal variation
ANNICK DE HOUWER: The acquisition of two languages from birth: a case study
LILIANE HAEGEMAN: Theory and description in generative syntax: a case study in West Flemish
A. E. BACKHOUSE: The lexical field of taste: a semantic study of Japanese taste terms

Earlier issues not listed are also available

THE LEXICAL FIELD OF TASTE

A SEMANTIC STUDY OF JAPANESE TASTE TERMS

A. E. BACKHOUSE
Australian National University

CAMBRIDGE
UNIVERSITY PRESS

CAMBRIDGE UNIVERSITY PRESS
Cambridge, New York, Melbourne, Madrid, Cape Town, Singapore, São Paulo

Cambridge University Press
The Edinburgh Building, Cambridge CB2 2RU, UK

Published in the United States of America by Cambridge University Press, New York

www.cambridge.org
Information on this title: www.cambridge.org/9780521445351

First published 1994
This digitally printed first paperback version 2005

A catalogue record for this publication is available from the British Library

Library of Congress Cataloguing in Publication data

Backhouse, A. E.
The lexical field of taste: a semantic study of Japanese taste terms / A. E. Backhouse.
p. cm. (Cambridge studies in linguistics. Supplementary volumes)
Includes bibliographical references and index.
ISBN 0 521 44535 3
1. Japanese language Semantics. 2. Taste Terminology.
I. Title. II. Series.
PL665.B33 1994
495.6'0143–dc20 93–38162 CIP

ISBN-13 978-0-521-44535-1 hardback
ISBN-10 0-521-44535-3 hardback

ISBN-13 978-0-521-02322-1 paperback
ISBN-10 0-521-02322-X paperback

For Kazuko

Contents

Preface xi

1 Taste and taste terms 1
1.1 Taste: narrow and broad perspectives 1
1.2 Taste terms 6
1.3 Taste and colour 12
1.4 Taste norms 16

2 Lexical meaning 19
2.1 Intralingual meaning 19
2.2 Extralingual meaning 22
2.3 Basic and extended meanings 26
2.4 Stylistic factors 27

3 Methodology 32
3.1 Delimitation and semantic analysis 32
3.2 Collection of data 34
3.3 Conventions of representation 35

4 Evaluative taste terms in Japanese 37
4.1 Delimitation 37
4.2 Grammatical properties 38
4.3 Syntagmatic relations 40
4.4 Paradigmatic relations 41
4.5 Stylistic factors 51
4.6 Extralingual meaning 62
4.7 Extended meanings 71

5 Descriptive taste terms in Japanese (I) 73
5.1 Delimitation 73
5.2 Grammatical properties 80
5.3 Semantic subdivision 84
5.4 The lexical system AJI I 88

6	**Descriptive taste terms in Japanese (II)**	141
6.1	The lexical system AJI II	141
6.2	The lexical system AJI III	142
6.3	AJI: the overall field	148
6.4	Extended meanings	151
7	**Retrospect and prospects**	160
7.1	Review of findings	160
7.2	Future directions	162
	Appendix: Summary of Japanese taste terms	173
	Notes	175
	References	187
	Index	193

Preface

The aim of this book is to provide a semantic description of taste terms in modern spoken Japanese. It originates in a doctoral dissertation presented to the University of Edinburgh in 1978, which has subsequently undergone substantial revision and expansion. The main additions in the present work take account of more recent publications in the area, provide additional sources of Japanese language data, and give wider consideration to cross-linguistic aspects of taste terms.

The book is organized in the following way. Chapters 1–3 are background chapters, dealing with the general study of taste and taste terms, lexical semantic fields, and the methodological approach adopted in the work. These are followed by the analysis of the Japanese terms: evaluative taste terms (glossed as 'good-tasting'/'bad-tasting') are described in Chapter 4, descriptive taste terms ('sweet', 'astringent', 'aromatic', etc.) in Chapters 5 and 6. The final chapter relates the findings to broader issues in the study of taste and taste terms, and offers some directions for further research. The Japanese terms described are brought together for convenience in the Appendix, where they are listed together with a brief summary of their semantic properties.

I am indebted to a number of people for their help and advice during the evolution of this study. Valuable comments on various stages of the work have been provided by John Christie, Bob Dixon, George Jelinek, J. V. Neustupný, and Anna Wierzbicka, and by three anonymous readers for Cambridge University Press. Portions of the analysis were presented in earlier versions at conferences and seminars in Brisbane, Canberra and Osaka, and I am grateful to the participants on those occasions for their comments. Sections of the final work were undertaken during research visits at Osaka University in 1989 and in the Research School of Pacific Studies, Australian National University, in 1991; my thanks go to the Departments concerned, as well as to the Australian National University for facilitating these periods of leave. I would like to thank Catherine Max at Cambridge for her patient guidance through the publication process, and to acknowledge the help of Panida Thamsongsana and Vivienne Kacsof in preparing the text. Finally, I must record my special thanks to my wife, who has not only provided continued support and encouragement but who also acted as the primary informant for this study.

1 *Taste and taste terms*

Our purpose in this first chapter is to provide relevant background information on the study of taste and taste terms. 1.1 presents a brief introduction to the scientific study of taste, and existing studies of taste terms are surveyed in 1.2. Compared with taste the neighbouring field of colour has been widely explored by both psychologists and linguists; in 1.3 we discuss some areas of contrast between the two fields and consider their possible linguistic implications. Finally in 1.4 we take up the specific topic of taste norms.

1.1 Taste: narrow and broad perspectives

This section is intended as a brief introduction to the work of physiologists, psychologists and food scientists in the study of taste.[1] A comprehensive survey of the field, apart from being outside the competence of the present author, is unnecessary for our purpose: our main emphasis is on the distinction between narrow and broad approaches to the study of taste, and on the implications of this distinction for linguistic studies.

In considering scientific work on taste one must begin by recognizing that the sense of taste is there generally defined in terms of a given set of sense organs or receptors, namely the taste cells. These are largely concentrated in thousands of taste buds found in papillae located predominantly on the tip, sides and back of the tongue; a small number of taste buds are also present in other sensitive parts of the mouth, including the pharynx and soft palate. The taste cells are stimulated by sapid (i.e. taste-bearing) substances when in solution or when dissolved in the saliva. Since the cells are sensitive to chemical energy, they are known as chemoreceptors, and the senses of taste and smell (whose chemoreceptors are the cells of the olfactory epithelium) are together referred to as the chemical senses. Scientifically, the study of taste is thus normally restricted to the study of various forms of sensitivity and perception arising from the stimulation of the taste cells by sapid substances in solution.

Compared with the senses of vision and hearing, the workings of the chemical senses are poorly understood. Work on taste has generally assumed that there are four primary tastes in man, namely sweet, sour, salty and bitter, and that these are detected by taste buds in different areas of the tongue and mouth. Scientific discussion continues to be conducted in terms of these four qualities (ostensively defined in terms of the chemical substances sucrose, tartaric acid, sodium chloride and quinine, respectively), but their status has frequently been called into question:

> The range of qualities is subdivided for convenience into the four common tastes, Sweet, Sour, Salt and Bitter...(They) are referred to here deliberately as the *common* tastes so as to avoid any assumption that they are genuine primaries. There is relatively little evidence to support the concept of primary tastes. Whereas attempts have been made to synthesize more complex tastes in terms of these four qualities, there are certain tastes, such as Metallic, which cannot be synthesized in this way. (Harper, 1972: 194)

Writing seven years later, McBurney and Gent (1979) conclude that available evidence does support the notion of four basic tastes (the metallic taste, for example, resulting from interaction with olfaction, i.e. smell); they also suggest that these four qualities should be regarded as separate senses or at least as sub-modalities, comparable to the various skin senses, a position first adopted by Öhrwall in the late nineteenth century (cf. Bartoshuk, 1978: 8–10). As for sensitivity to these four qualities, it is now accepted that different areas of the tongue and mouth show variations in sensitivity to a given quality, the tongue tip, for example, being most sensitive to sweet, but the view that individual receptors are uniquely sensitive to a particular quality has been abandoned.

The precise nature of the stimuli involved in the action of the chemical senses is still not known. It is clear, however, that there is no simple correspondence between chemical composition and perceived taste quality: salts, for example, do not all taste salty, nor do all acids taste sour; substances of various chemical groups (including, for example, dilute sodium chloride solutions) share a sweet taste, and there is often a close chemical association between sweet and bitter. Experimental data on the perceived taste of mixtures of the four common qualities indicates that more than a simple additive process is involved: it is often impossible to predict the taste of a mixture from the taste of the separate constituents. Moreover, the results of such tests tend to vary across subjects, and individual differences in sensitivity are widely acknowledged. The internal state of the body is recognized as one important factor here: salt deprivation, for instance, increases sensitivity to salty tastes, while water deprivation decreases it. In addition, general taste sensitivity tends to decrease with age, though here again individuals differ considerably. Genetic differences may also be involved, and a kind of taste blindness is also known to exist with certain substances: phenylthiourea, for example, tastes highly bitter to some people, but is virtually tasteless to others.

Several psychological scales of taste have been devised. Absolute threshold concentrations for different taste stimuli vary widely: it is possible, for instance, to detect a salt solution half the strength of the standard sugar solution. Here again there are many factors which cause variation in sensitivity, notably the temperature of the taste stimulus and, as indicated above, the area of the tongue stimulated. It is noticeable that the majority of modern foods (which are almost all taste mixtures) obtain low rankings on taste scales, exceptions being substances such as vinegar (perceived as highly sour) and honey (highly sweet), which are not normally consumed alone. Different taste stimuli also give rise to variable sensations of pleasantness or unpleasantness according to concentration: in particular, whereas bitter, sour and salty substances are quickly felt as unpleasant as concentration increases, the pleasantness of a sweet substance such as sucrose is maintained at a high level.

This points to the fact that, like all senses, the sense of taste to some degree fulfils a biological alerting function, with substances harmful to the human organism being perceived as unpleasant. Thus bitter is an important indicator of toxicity, and Wyburn et al. speak of 'an almost universal aversion to bitter tastes' (1964: 122). By contrast, sweetness in ripe fruit is an important indicator of high vitamin C content and is in general associated with high calorie substances; similarly, human beings have a specific hunger for sodium, and salty foods, as we have seen, are generally perceived as pleasant at low concentrations (Goldstein, 1984: 422, 425). It has also been noted that neonates show clearly distinguished facial reflexes towards sweet, sour and bitter taste stimuli (cf. Moskowitz, 1978: 176–7). However, the correlation between affectivity and nutritional value is not perfect, and in the case of humans biological indications may be overruled by cultural influences, as illustrated, for example, by the existence of 'acquired tastes' in various societies: these involve substances which are generally perceived initially as unpleasant but whose consumption is nevertheless cultivated in the culture concerned; they are persevered with on this basis, and eventually come to be perceived as pleasant.

Many scientists have pointed out that the narrow definition of taste outlined above does violence to everyday usage, and have urged that scientific work should be based more upon the folk concept of taste. As long ago as 1936, Rubin proposed 'a scientific definition of "taste"...in accordance with popular use' and recommended that 'the starting-point for every discussion of taste must be the experience of daily life' (1936: 74). In contrast to the prevailing scientific division into four taste qualities, he quotes with approval the earlier ten-fold scheme put forward by Linnaeus (aqueous, acid, fatty, sweet, mucous, dry, bitter, astringent, acrid/sharp, salt), his only reservation being that even this classification may distinguish too few categories (1936: ibid.). This may be compared with Aristotle's eight-fold scheme of sweet, oily, pungent, rough, astringent, sharp, salt, bitter (Hamlyn, 1968: 38).[2] More recently, Gibson has contrasted the everyday sense of taste as the whole perceptual system accompanying eating with its scientific definition which reduces it to 'a minor sense, for distinguishing solutions' (1966: 136). Common to these objections is the view that, in everyday usage, taste covers a much wider range of qualities than those perceived through the taste cells alone.

In the interests of clarity, we propose to adopt from this point on the terminological distinctions suggested by Rubin (1936: 78) and commonly employed elsewhere: the term 'gustatory' will be applied to those qualities perceived via the taste cells (i.e. to taste qualities in the narrow, scientific sense), and the term 'olfactory' to those qualities which arise from the olfactory epithelium. The term 'taste' is thus free to be retained in its wider, everyday sense.

It is a well-known fact that many qualities which are subsumed under taste in everyday usage are in fact olfactory. Everyone knows how a cold in the head dulls taste sensitivity, and it has been shown experimentally that, when the nose is held closed, it is in general impossible to distinguish between, for example, the tastes of raw apple and raw onion. Flavours such as vanilla are also not tasted unless the nose is released. The olfactory epithelium is stimulated by volatile substances, either directly via the nostrils or by way of the back of the mouth. Where food is involved, it is therefore sensitive to stimulation both before and during eating, and smell acts as an important indicator of the soundness of food.

In everyday usage we also speak of 'sharp' tastes (for instance, with reference to fizzy drinks or spicy foods). In this case it appears that neither the gustatory nor the olfactory receptors are responsible, but receptors analogous to the pain and temperature receptors in the skin and as yet poorly understood. Gibson (1966: 139) comments on such substances as follows: 'Mustard seems to affect the skin as heat does, and menthol as cold does, while pepper and horseradish yield something like tingling pain.' He goes on to note that adults, but not children, cultivate such non-nutritive condiments: they are cases of 'acquired tastes'. Another quality subsumed under taste in common usage is astringency, which also appears to involve some form of cutaneous stimulation. Harper (1972: 279) comments thus: 'To the four classical subdivisions of cutaneous sensitivity...should also be added Pungency as a tactile accompaniment of many odour stimuli and Astringency which accompanies certain taste stimuli (e.g. strong tea without milk, or rhubarb).'

Rubin maintains that, in ordinary language, 'taste' applies to 'the properties of food as they appear when the food is placed in the mouth and masticated' (1936: 77) and accordingly goes on to mention many other types of perceptual qualities involved in the taste experience: being an area of high tactile sensitivity, the mouth yields information relating to such qualities as texture, consistency, shape and temperature; hearing, too, as when one crunches an apple or a biscuit, enters into the perception of the consistency of food and of the modification of this consistency during mastication. Moreover, he emphasizes that the taste experience is not merely the static cooperation of all the sense organs, known and unknown, involved, but a sequential process continually changing through time and including (for solids) the stages of biting, mastication, insalivation and swallowing.[3] Gibson, as we have seen, stresses the everyday meaning of taste as a perceptual system accompanying eating and uniting several different types of receptors in this single function, and he presents a range of information perceived by the system similar to Rubin's list (1966: 138). (Cf. also Amerine et al., 1965, Chapter 4.)

The points raised by these authors clearly indicate that a linguistic study of taste terms in everyday language must be prepared to find that their denotational range takes in dimensions other than gustation alone. The above discussion suggests, for example, that in addition to terms relating primarily to gustation, terms denoting olfactory and various cutaneous and tactile 'taste-feel' qualities are subsumed under taste in everyday English usage.[4] Just how far the denotation of the verb or noun TASTE in everyday English ranges over the whole ingestion experience is of course an empirical question to be decided by linguistic investigation: part of the evidence to be considered will be the relative acceptability of words such as SWEET, SMOKY, SHARP, CRUNCHY, LUMPY, etc., when used in answers to such questions as *What does X taste like?* Furthermore, as with any lexical field, the precise denotational domain of taste terms is likely to vary across languages.

1.2 Taste terms

In this section we introduce existing studies of taste terms, chiefly by linguists or anthropologists, but also by taste scientists where these have relevance for everyday language.

In general, the linguistic study of taste terms has received much less attention than that of colour terms; one reason for this is no doubt the lack of scientific understanding of taste that has been noted in the previous section, and the consequent absence of any established psychophysical framework comparable to that which exists for colour (cf. Sturtevant, 1964: 103). Until relatively recently, it was necessary to go back to the first decade of this century to find studies directly concerned with the language of taste, and we begin by considering these.

In this period, interest focussed on investigations of the sense vocabularies of 'primitive' peoples. Chamberlain's study (1903) is a survey of taste terms in the Algonquian languages: Algonquian terms are listed and discussed under various English taste categories, information supplied by the author being supplemented from lexicographical sources. Discussion ranges over 'generic' taste terms (terms corresponding to the English noun and verb TASTE, also terms meaning 'good-tasting', 'bad-tasting' and 'insipid, tasteless') and 'special' taste terms, mainly under gustatory headings but also including such categories as 'astringent', 'pungent' and 'rancid'. The evidence as presented indicates the widespread presence of distinct terms for 'sweet', 'sour' and 'bitter', with the term for 'sour' being commonly extended, either directly or by way of derivation, to cover salty tastes.

In Myers (1904), the author, a psychologist, begins by summarizing his and Seligmann's efforts to investigate the taste vocabulary of the Torres Strait islanders. The investigators tested the islanders with dilute solutions of sugar, salt, acid and quinine, 'and thereby obtained their equivalent words or expressions for sweet, salt, sour and bitter' (1904: 118–19). Myers' summary of the findings is as follows:

(i) The literal meaning of the phrase commonly used in the Torres Straits[5] to denote sweetness is 'tasting good'.
(ii) The same phrase is applicable to denote saltness.
(iii) The usual word for saltness is derived from the word for sea-water.
(iv) The taste-names for salt and sour tend to be confused.
(v) There is no specific name for the bitter taste. (1904: 119)

Myers followed up these experiments by sending questionnaires to officials, missionaries and other individuals in contact with primitive peoples in various parts of the world. The relevant questions may be summarized as:

(i) By what words in their own language would the natives describe the taste of solutions of (a) sugar, (b) salt, (c) *weak* acid, (d) quinine? (Where possible, investigators should carry out tests using such solutions.)

(ii) Give, if possible, the exact meaning of the words in (i). (E.g. Are they derived from substance names? Are they extended uses of words, such as 'soft', 'sharp', or have they given rise to such? Do they mean 'pleasant' and 'unpleasant'?, etc.) (Cf. 1904: 121)

The results of this questionnaire he sums up as follows:

(i) Several languages have two taste words, one applied to sweet, salt and other agreeable flavours, the other to unpleasant flavours.

(ii) The use of a common word applied to both sweet and salt is widely attested.

(iii) The taste word for salt, where present, is commonly derived from the word for sea-water.

(iv) Confusion between salt and sour, and more especially between salt and bitter, is frequently attested. In New Guinea, New Hebrides, and much of Polynesia, the same word denotes salt, sour and bitter. These words doubtless mean 'rough', 'unpleasant' or 'biting'.

(v) In several languages, the same word denotes sour or bitter. A notion of distaste, astringency or pain underlies these words. (Cf. 1904: 122–4)

Myers' general conclusion, once more, is that in primitive languages taste vocabularies, like colour vocabularies, show a lack of differentiation when compared with English, and, in the most primitive, 'substances are primarily classed according as they are tasteful or distasteful; the differences, for example, between sour and bitter are considered less striking than their common impalatability' (1904: 126).

Myers' findings clearly need to be evaluated with care. Thus, the methodology of the original experiments may be questioned, relying as it does on an ostensive 'look-and-say' technique: one possibility here, for example, is that the islanders tended to respond to the stimuli at the general evaluative level of 'pleasant'/ 'unpleasant' and that more specific descriptive terms, though existing, were not elicited in the context of the experiment. In addition, it could be claimed that the stimuli employed were unnatural and unsuited to cross-cultural studies of this kind. The same objections apply to the questionnaire; here moreover, as Myers is aware, the validity of the results will clearly depend in large measure on the degree of familiarity with the native language on the part of the intermediaries who carried out the tests. Finally it follows from the considerations raised in the previous section that, even if the results are accepted as being accurate as far as they go, an investigation restricted to the four gustatory qualities can give only a limited indication of the overall structure of the taste vocabularies of the languages concerned.

Despite these shortcomings, Myers' investigations represent a valuable attempt to extend cross-linguistic studies of the vocabulary of perception to taste terms. He also draws attention to parallels between 'primitive' and Indo-European languages in the vocabulary of taste. Particularly interesting are cases for which he cites commonly associated substances: Greek PIKRÓS, for example, he cites as being applied to sea-water, unripe fruit and pungent flavours, while in Sanskrit (which is said to have recognized six tastes: sweet, salt, sour, bitter, pungent, astringent) the corresponding term TIKTA was applied to gourd, mustard and other burning tastes; outside Indo-European, Somali DANAN is cited as being applied to lemons and salted meat (1904: 120, 124).

Finally from this period, Rivers (1905) contains a brief section dealing with the taste terms of the Todas of southern India. Like Myers, Rivers made use of solutions aimed at eliciting terms for the four gustatory qualities. Apart from reporting some difficulty in administering the solutions ('as the Todas refused to allow me to put the solutions into their mouths'), Rivers records that distinct terms were found for all four qualities, and that the only confusion of tastes occurred between salt and sour (1905: 388).

Following these early studies, taste terms were ignored by linguists and anthropologists until relatively recent times. In English, Lehrer (1975, 1983) has discussed the use of English terms in the description and evaluation of wine: she focusses in particular on the role of the intralingual structure of the vocabulary in facilitating the extension of terms to this domain, and on the degree of consensus in the way different speakers apply terms to wines. The mechanism of semantic extension is also explored in Lehrer (1978), taking terms applied in the wine domain as an illustrative set. Lehrer's work shows clearly that the English wine vocabulary is particularly vast and that conversation about wine shares many of the characteristics of aesthetic discourse. Kuipers (1984) discusses a smaller set of taste terms in Weyéwa, an Austronesian language of Sumba, in eastern Indonesia. He emphasizes the distinction between scientific and folk concepts of taste, and proposes, on linguistic and cultural grounds, a set of seven terms (sour, sweet, salty, bitter, tart, pungent, bland) as basic taste terms for the language. Kuipers is centrally concerned with the use of taste terms in discourse, and he points out that in everyday conversation it is relatively unusual to use taste terms to refer to actual taste sensations; by contrast, the Weyéwa terms for bitter and bland occur as a common opposition in ritual speech, where they are extended (broadly with the meanings 'prohibited' (bitter)/'permitted' (bland)) to the domain of ritual status. Apart from these studies specifically concerned with the language of taste, Sturtevant's earlier survey of ethnoscience mentions the domains of smell and taste as having, in English, a small and weakly terminologized vocabulary, and confirms that, in comparison with colour, the psychophysical dimensions involved are very poorly known (1964: 119). He refers to a brief study by Aschmann published in 1946 of the smell categories of Totonac: as reported by Sturtevant, Aschmann's analysis shows a taxonomy with eight primary terms each dominating more specific terms; because of the lack of an established classificatory scheme, the range of application of each term is exemplified by listing words denoting objects which are typically characterized by the smell concerned. (The primary terms are also discussed in Nida, 1949: 158–9.)

Linguistic and anthropological work published in Japanese centres on the taste terms of the standard language and of Japanese dialects, but also includes some studies of taste terms in other languages. The primary descriptive treatment of the standard language is contained in Nishio (1972: 98–105, 272–4, 327–9), a comprehensive survey of the meaning and use of Japanese adjectives exemplified largely from a literary corpus. The main additions to this treatment in Nishio (1983) are arguments attempting to establish certain terms as basic in the taste domain. Discussions of taste terms in Japanese dialects date from the publication of the Linguistic Atlas of Japan (Kokuritsu Kokugo Kenkyuujo, 1966–74), which investigated terms for sweet, salty, hot, sour and good-tasting, revealing considerable geographical variation. The findings for these terms were subsequently discussed in articles by Katoo (1966), Fujisawa (1975) and Nagao (1982, 1983); Nagao's studies also deal with the historical development of Japanese taste terms, as does Hachiya (1983). Yamada (1972) and Kunihiro (1982: 150–3) deal with the taste term systems of the Miyakojima (Okinawa) and Yamaguchi dialects, respectively, and contrast these with Tokyo Japanese.

Maeda (1978) is a description of taste terms in modern Korean, based on linguistic investigation with informants. Maeda reports that, as in other areas of the vocabulary of perception, Korean is richly endowed with taste terms. These centre on seven items (sweet, sour, salty, appropriately salty, bitter, hot, astringent), from which further terms are derived by suffixation; these derivatives, in turn, produce further forms through processes of vocalic and consonantal alternation, resulting in a total of around ninety terms. Maeda examines the semantic effects of these phonological processes, relating vocalic alternation to taste intensity, taste purity (vs. admixture) and palatability, and consonantal alternation to intensity. From his results, it appears that the function of vocalic alternation is primarily affective. This accounts for the special linguistic status of terms based on 'sweet' brought out in the analysis: the same alternation indicating high intensity and purity in the case of 'sweet' denotes low intensity and purity in the case of other terms, and this may be interpreted as a linguistic reflex of the inherent palatability of the sweet taste.[6]

Ishige (1983) is a wide-ranging article on biological, cultural and linguistic aspects of taste written by an anthropologist with a special interest in this area. While pointing to the biological basis of the sense of taste, reflecting, for example, man's nutritional need for sweet substances and (in appropriate quantities) for salt, Ishige stresses the importance of the cultural component in the taste preferences of human societies and the need to investigate the structure of taste vocabularies as indicators of cultural patterns. As illustrations, he discusses the taste terms of Ponapean, Galelan (a Papuan language of Halmahera, Moluccas: based on Ishige, 1978) and Anatolian Turkish (based on Matsubara, 1976). Ishige reports that Ponapean possesses a basic opposition between 'sweet' and 'bitter', correlating with 'ripe' and 'unripe' as applied to items of the staple Ponapean diet. Sweet tastes are characterized as good-tasting, except in the case of the narcotic drink *sakau* (kava), where bitterness is prized (and correlated with ripeness): this is clearly an excellent example of an 'acquired taste'. Galelan possesses terms for 'sweet', 'salty', 'sour', 'hot' and 'bitter'. For items to qualify as good-tasting, these last four qualities must be present in exactly appropriate amounts: both 'too salty' and 'not salty enough', for example, are expressions of unpalatability. In the case of 'sweet', however, only 'not sweet enough' falls into this category, and items cannot be too sweet to be palatable. Interesting features of Anatolian Turkish taste terms reported by Ishige include the presence of a superordinate term ('harsh, pungent') ranging over terms for 'sour', 'salty', 'hot', 'acid' and 'bitter', and the expression of both 'sweet' and 'good-tasting' by a single term.

Amongst work by scientists with linguistic relevance, Robinson (1970) notes that the application of the terms BITTER and SOUR in everyday English usage differs considerably from their scientific definition. In tests with untrained observers, 67 per cent characterized the taste of lemons as sour, 33 per cent as bitter, while in the case of lime juice 39 per cent judged it as sour and 39 per cent as bitter. Robinson's conclusion is that, since few food substances taste predominantly bitter (in the scientific sense), it is possible to go through life without clearly identifying the taste and that, in view of its unfavourable connotations, the tendency may be to use the term for unpleasant tastes, particularly strong sour tastes. (Lehrer (1975: 912–13) notes a high correlation in the use or non-use of both these terms with reference to a given wine: i.e. subjects tended to apply both terms together, or to use neither. Her suggestion is that the unpleasant taste (for which English has no general term) is of primary importance, rather than the further distinction between 'sour' and 'bitter'.) Harper's survey (1972) stresses the need for linguistic studies in the lesser understood areas of perception, including taste: amongst other problems, he points out that the extent to which different people agree in their understanding and use of everyday descriptive terms needs to be examined, and he adds that 'dictionary definitions are rarely satisfactory in this context, since they are not anchored to particular stimuli' (1972: 38).

Finally, the work of O'Mahony and his colleagues focusses on descriptions of taste stimuli given by speakers of English and speakers of other languages: O'Mahony and Muhiudeen (1977) examines English and Malay, O'Mahony and Tsang (1980) English and Cantonese, O'Mahony and Manzano Alba (1980) English and (Mexican) Spanish, and O'Mahony and Ishii (1986) English and Japanese. Results for Malay, involving descriptions of salt solutions, indicate that Malay speakers often make use of complex expressions such as MASIN AYER LAUT 'salty like sea-water', MASIN GARAM 'salty like salt', MASIN KITCHUP 'salty like soy sauce' and MASIN MAUNG 'salty, obnoxious' in discriminating tastes (although MASIN 'salty' was also commonly used alone); similar expressions are reported to occur with other taste qualities, such as MANIS BUAH 'sweet like fruit' and MANIS GULA 'sweet like sugar'. Experiments for the other languages involved a wider range of taste stimuli. In the case of Cantonese, there was a tendency for Cantonese speakers to describe the taste of monosodium glutamate (widely used in food preparation in East Asia) in specific terms, whereas English speakers commonly used SALTY, either alone or in combination with other terms. A similar trend was observed for Japanese, and these results lead O'Mahony and his co-authors to speculate that, had early taste research been performed in East Asia, a 'glutamic' taste might well have been proposed as an additional primary. The main point of interest in the Spanish study concerns the existence of three terms (ÁCIDO, AGRIO, AMARGO) in the sour/bitter area. As well as responding to particular taste stimuli, subjects were asked to answer a questionnaire matching taste terms with a list of foods: while the presentation of the results of this study is marred by typographical problems in the tables, it appears that, for monolingual Spanish speakers, AMARGO is commonly applied to the taste of quinine sulphate, and associated with strong black coffee and beer; similarly, ÁCIDO is commonly used to describe the taste of citric acid, and associated with limes, vinegar and unripe fruits; for AGRIO, the clearest result is a common association with orange peel, and the authors conclude that the status of this term, inconsistently described in dictionaries, remains unclear. There was some evidence of ÁCIDO/AMARGO confusion. In this connection the authors cite Malay as possessing a large range of terms in this area (MASAM 'sour', KELAT 'acidic', PEDAR/PAYAU/PAHIT 'bitter'), and note the potential value of the language for the study of taste confusions.

Relevant details of the works cited in this section will be discussed further at appropriate points in the study.

1.3 Taste and colour

Of all sense vocabularies, it is colour terms which have attracted most attention from semanticists (cf. Sturtevant, 1964; Berlin and Kay, 1969; Kay and McDaniel, 1978; Wierzbicka, 1990; and the references cited therein), and they provide a natural point of comparison for studies of terms in neighbouring perceptual fields. Taste and colour vocabularies will be compared and contrasted at various points throughout this study. Here, by way of introduction, certain characteristics of taste perception are informally compared with aspects of colour perception, and some possible linguistic implications considered.

The first point is that taste and colour perception differ in complexity in terms of the variety of psychophysical parameters and receptors involved: it seems clear that, except in the narrow scientific sense, taste ranges over at least a good part of the eating and drinking experience, and involves several different systems of receptors. It may be that this relative complexity will be reflected in the structure of taste vocabularies, with, for instance, a number of distinct lexical systems each relating to different parameters involved (though, of course, not necessarily in one-to-one correspondence with the scientific divisions set up by psychologists). In this connection, it is worth bearing in mind that, of the analytically distinct parameters concerned (gustation, olfaction, and various cutaneous and tactile phenomena such as astringency, pungency, temperature, consistency, texture and hardness), gustation and, perhaps, astringency and pungency are peculiarly unique to the sense of taste; it is possible that the terms used may reflect this difference, in terms, for instance, of the basic or extended status of their meanings.

Secondly, colour perception, as an aspect of vision, and taste perception contrast in terms of distal vs. proximate receptors: vision and hearing are commonly grouped together as the distal senses, gustation, olfaction and touch as the proximate senses. It is commonly assumed that the vocabulary associated with the proximate senses is in some way poorer than that associated with the distal senses: Sturtevant's comments on smell and taste (i.e. gustation) terms in English have already been noted, and Slobin maintains that we have 'an inadequate vocabulary for expressing sensations of the proximity senses' (1971: 108). Such statements are to be interpreted as referring to basic terms, since Lehrer's work shows clearly that, at least for the description of wine in English, any shortage of basic vocabulary in this area is readily offset by means of word-formation and, especially, by semantic extension from other domains: 'Though it may be true that only a small number of words are used exclusively or primarily for tastes, there are in fact dozens of terms that can be and have been used to describe and evaluate wines' (1975: 902; cf also 1983: 16–18, 129–35). Detailed linguistic studies of the vocabularies concerned can enable us to investigate the precise nature of this putative 'poverty' and determine how far it holds across different languages.

A third property of taste perception that sets it off from colour perception, and perhaps from the operations of all the other senses, is that it is par excellence a 'cultural' sense. By this we mean that taste perception is geared to the activities of eating and drinking which, in humans, are pre-eminently culturally channelled activities: except in infancy, when the mouth plays an important part in a baby's exploration of the world around him,[7] the input to taste perception is culturally preselected in that what is accepted as food, or 'good to eat', in a given community is in general rigidly determined in a way that does not apply in the case of the other senses. Taste terms are predicated of items of diet, and not indiscriminately of objects in the environment. Since this is so, and since the contents of diet and methods of cuisine vary considerably across cultures, we must expect a good deal of cross-linguistic variation, both in the nature of the dietary items with which taste terms are typically associated, and in the particular aspects of the taste experience which are lexically emphasized. As with any area of the vocabulary, we must expect the elaboration of taste vocabularies to reflect the interests of the speakers of the language concerned: since the taste experience is a complex one, languages may well differ, both in the overall attention they pay to the language of taste, and to the selection of particular areas within it for special emphasis.

A fourth, major difference between colour and taste perception is that the latter clearly involves an affective aspect: at a general level, some tastes are inherently pleasant, some inherently unpleasant (cf. 1.1). This is often reflected in dictionary definitions of the corresponding terms: definitions of SWEET as 'having or denoting a *pleasant* taste like that of sugar' and of BITTER as 'having or denoting an *unpalatable* harsh taste, as the peel of an orange or coffee dregs' (Hanks, 1986: italics mine) contrast markedly in this respect with definitions of colour terms. (Importantly, general affective properties of this kind can of course be 'overruled' by the taste expectation for a particular substance: sweet beer, for example, would not normally be considered pleasant despite the generally favourable affect of sweetness. Cf. 1.4 below.) In view of the biological alerting function ascribed to gustation, there seems no reason to doubt that this is a universal feature of this domain. Indeed, similar considerations no doubt apply to the other proximate senses, so that affective value seems likely to be a pervasive feature of taste terms in the wider sense: cf., in this connection, the repeated emphasis on the evaluative content of wine words in Lehrer, 1983. Linguistically, factors of this kind will be reflected in, for example, the syntagmatic semantic relations of the terms concerned.

A fifth, no doubt related, difference between colour and taste perception is that the latter is characterized by a much greater degree of individual variation. This is most apparent in the case of explicitly evaluative terms such as GOOD, DELICIOUS, etc.: one man's meat is often another man's poison, and we naturally expect disagreements among different speakers in their application of such terms to given substances. Even in the case of more descriptive terms, however, apart from the experimental evidence of individual differences reported in 1.1, everyday experience teaches us that people vary considerably, for example, in the amount of sugar they take in their tea and in their sensitivity to acquired tastes such as the hotness of spicy foods and the bitterness and astringency of certain alcoholic drinks, so that what is sweet tea, a hot curry, or a dry wine for one individual may not be so for another. This would seem to have no general parallel in the field of colour: in general, it appears to be a fact that members of a cultural group are prepared to accept greater subjectivity in taste judgements than in colour judgements. Linguistically, we will expect such disagreements to be reflected in the application of taste terms to given items of diet, as indicated above, rather than in differences in the central intralingual structure of the lexical system itself (cf. Lehrer 1983: 85, 94–6).

The above discussion has highlighted certain contrasts between colour and taste perception, which in turn point to possible semantic differences between colour and taste terms. To conclude this section, we note differences in the linguistic status of colour and taste terms suggested by Dixon's (1982) cross-linguistic survey of adjectives.

Dixon's central concern is to investigate how terms from semantic domains which belong to the adjective class in English are expressed in other languages, some of which have only a small, closed adjective class or lack one entirely. To this end he postulates a series of universal semantic types, including 'colour' and 'physical property', descriptive taste terms such as English SWEET and SOUR being assigned to this latter together with adjectives of weight, texture, temperature, etc. (Evaluative taste terms such as DELICIOUS are included in 'value'; we shall ignore these here.) The separation of 'colour' and 'physical property' types is justified on a variety of semantic, syntactic and morphological grounds in English. Thus, colour terms are structured intralingually as an incompatible set, and also exhibit relations of hyponymy; comparative forms of colour adjectives (*redder*, etc.) are semantically quite unlike the comparative forms of other adjectives; by contrast, most physical property adjectives form antonymous pairs, although the taste terms SWEET, SOUR, SALTY and BITTER are also noted as sharing some characteristics of incompatibles (cf. 1982: 19). Syntactically, English adjectives from these types occupy different slots in noun phrase structure, physical property terms occurring earlier than colour terms, with 'speed', 'human propensity' and 'age' types intervening between the two (1982: 24): predicted normal orderings are thus illustrated by *sweet young port* and *old brown kipper*. Morphologically, all physical property adjectives, but only some colour adjectives, form derived adverbs in -LY (BLACKLY, SWEETLY, SOURLY, BITTERLY); semantically, these adverbs generally exhibit extended meanings of the 'human propensity' type (cf. 1982: 31). In general, the 'colour'/'physical property' division established for English is supported by data from other languages in that the two types often differ in word-class membership: in particular, colour terms are likely to belong to the adjective class, however small, whereas physical property terms frequently appear as verbs in languages with few adjectives (1982: 46–7). The chief exception here is Telugu, where the four colour terms and the four taste terms belong, with a handful of other items, in the same special closed class of roots (1982: 42), manifesting, for one language at least, a close affinity between the two domains. Dixon notes this, but also underlines its exceptional status in the languages surveyed (1982: 49, note 54): 'if we worked intensively with Telugu...we might well be led to set up a semantic type involving "colour" and "taste". (This would not, however, be confirmed by data from other languages.)'

1.4 Taste norms

In 2.2 we point to the importance for extralingual semantics of verbal indicators of focal exemplars of denotational categories: we suggest that generic propositions expressed by sentences incorporating focal exemplars, such as *Blood is red*, can be regarded as denotational axioms. We further suggest that speakers will tend to agree on the choice of such axioms. While propositions of this kind are normally taken as unspoken 'givens' in everyday communication, they may be expressed in special situations (for example, when speaking to young children) and are indeed frequently drawn upon in dictionary definitions. As well as being worthy of study in their own right, such axioms appear to offer the only practicable approach to the study of denotation in many domains, including taste: as we have seen, not only has scientific work on taste so far failed to yield any established psychophysical scheme for the area, but psychologists have called for studies of the everyday language of taste as a necessary step in developing scientific understanding.

Axioms, then, have a valid place in our investigations, and examples from the domain of taste are readily found, both in English and Japanese. *Sugar is sweet*, for example, has its Japanese counterpart in *Satoo wa amai*: this is indeed an axiom which we would expect to hold widely across languages. Other familiar combinations are *Remon wa suppai* 'Lemons are sour' and *Kareeko wa karai* 'Curry powder is hot'. On the other hand, an example of a more 'exotic' axiom, at least from the viewpoint of English, is *Shibugaki wa shibui* '*Shibugaki* (a type of persimmon) are astringent'. Further exemplification and discussion of axioms will occur at relevant points in the semantic analysis.

In this section, we wish to distinguish an additional type of generic proposition, which appears to be highly characteristic of taste terms as a group. This type of proposition has its origin in the fact that food substances in general have associated with them expectations in terms of which, typically, they are recognized as 'good to eat': we shall refer to such propositions as taste norms, in contradistinction to the axioms which indicate exemplars of taste qualities. Examples of taste norms from Japanese are illustrated by such sentences as *Nori wa kaori ga ii* 'Laver is aromatic', *Toofu wa shitazawari ga ii* 'Bean curd is pleasantly smooth to the tongue', and *Mikan wa amazuppai* 'Mandarins are sweet-sour'.

Although there may be individual borderline cases, in principle axioms and taste norms are quite different in nature. Axioms serve to indicate food substances which centrally exemplify given taste qualities, thereby providing anchoring points for the application of taste terms; they answer the question *What tastes X?* Taste norms, on the other hand, indicate taste qualities which given food substances are normally expected to possess; they answer the question *What should X taste like?* As a general rule, therefore, terms denoting unpleasant taste qualities will not occur in taste norms. And whereas axioms are in general free from exceptions (non-sweet sugar, or non-sour lemons, lie outside the range of our normal experience), taste norms express expectations which may be broken in inferior instances of the food substance concerned. Thus, despite the existence of taste norms for laver, bean curd and mandarins referred to above, it is clearly possible to have instances of stale laver, of poorly refined bean curd, and of unripe or insipid mandarins which do not meet the expectations expressed. The existence of built-in expectations of taste is reflected in the fact that specific statements of the form *This X is Y*, where X denotes a food substance and Y is a descriptive taste term other than the taste norm(s) for that substance, are commonly to be interpreted not simply as plain statements of fact but as expressions of negative value judgement: in the case of mandarins, for example, the statement in Japanese *Kono mikan wa suppai* 'This mandarin is sour' (the taste norm being *Mikan wa amazuppai*, as we have seen, where AMAZUPPAI implies the presence of sweetness) is generally to be understood as indicating the speaker's disapproval. It is important to note that this is not necessarily a matter of the taste term itself, here SUPPAI 'sour', denoting an inherently unpleasant quality: pickled plums (*umeboshi*) are sour, but are eaten regularly (normally accompanied by rice). That a thing is sour is not enough to make it 'good to eat' or otherwise: what matters is whether it is supposed to be sour or not. This applies equally to the inherently pleasant quality of sweetness, which is out of place, for example, in many varieties of table wine (cf. 1.3). A mature speaker of a language has normally been conditioned – even to the extent of 'acquiring' what are unpleasant tastes in general biological terms (cf. 1.1) – to the dietary norms of his own culture, with respect both to what items are regarded as food and to why. As a result, personal liking in this area is no doubt governed to a considerable degree by whether cultural taste expectations are met in particular instances.

The domain of taste, then, is characterized by the existence of built-in taste expectations for food substances, and these may be verbalized more or less readily in different cultures as taste norms. In the case of Japanese, taste norms are readily elicited for many food substances and, like axioms, we shall record them as an important type of semantic data in this domain.

Impressionistically, at least, gastronomic matters attract a relatively high degree of cultural preoccupation in modern Japan. The Japanese are conscious of the culinary distinctiveness of their traditional diet (as opposed to the imported Chinese and Western sectors), a view generally shared by visitors to the country. Traditional Japanese food tends to strike the newcomer as somewhat insipid, and is commonly characterized as highlighting finer discriminations of flavour and texture within a somewhat restricted gustatory range.[8] This is perhaps reflected in the vocabulary relating to certain aspects of the taste experience where Japanese possesses terms for which it is very difficult to find English counterparts: as we have seen, bean curd is expected to be SHITAZAWARI GA II 'pleasantly smooth to the tongue', and a variety of further terms relating to texture and consistency are in common use.[9] How Japanese taste terms are organized in relation to the more familiar parameters of taste will be explored in the course of the semantic analysis in Chapters 5 and 6.

2 *Lexical meaning*

In this chapter we outline the approach to lexical meaning adopted in this study. This approach is basically that of the lexical semantic field, in particular as developed by Lyons in his work on lexical semantics (cf. Lyons, 1963; 1968: 400–81; 1977: 174–335). Broadly, the meaning of an item of the vocabulary is seen as being specifiable in terms of the various types of intralingual semantic relations which the item contracts with other items in the language, and in terms of its applicability to phenomena outside language (extralingual meaning).

The scheme described below constitutes the basic framework for the subsequent semantic analysis of Japanese taste terms. Intralingual and extralingual meaning are discussed in 2.1 and 2.2, respectively, while 2.3 briefly takes up some aspects of basic and extended meanings from the viewpoint of the present framework. 2.4 discusses the treatment of stylistic factors in the vocabulary.

2.1 Intralingual meaning

By intralingual meaning we refer to semantic relations which hold between lexical items in a given language.[1]

2.1.1 Paradigmatic relations

Here we are concerned with semantic relations which hold between lexical items that can occur in the same linguistic context.[2] Paradigmatic relations in general hold for a particular range of such contexts, rather than for the language as a whole.

Relations of this type rest upon the notion of implication holding between sentences containing the lexical items concerned. As a simple example, a sincere utterance of the sentence *This apple is sweet* will be held by speakers of English to imply the denial of the sentence *This apple is sour*, while no implication of any kind will follow with regard to, say, the sentence *This apple is crisp*. That is to say, part of the meaning of SWEET[3] in such contexts is its implicit exclusion of

SOUR, and these two lexical items are here paradigmatically related whereas SWEET and CRISP are not. Note that this relation between SWEET and SOUR does not hold in all contexts: in *She is a very sweet person* or *That was sweet of him*, for example, SWEET contrasts rather with items such as NASTY or UNKIND.

Lexical items which are paradigmatically related in this way are thereby said to belong to the same lexical semantic system.[4] Thus, in contexts such as the first example above, SWEET and SOUR belong to one lexical system, while SWEET and CRISP do not. Membership of a given lexical semantic system is part of the meaning of a lexical item. According to the nature of the implications that hold, various types of paradigmatic relations between lexical items can be distinguished, and the lexical system concerned is said to be structured in terms of them. The types of relations found, and the resultant types of structuring that exist in lexical systems, are important areas for empirical investigation within and across languages.

Broadly, paradigmatic relations involve inclusion (with identity as a special case), or exclusion. Among the types of relations that have been distinguished are the following: synonymy (e.g. GLASSES/SPECTACLES), hyponymy (TREE/ELM), partonymy or meronymy (LADDER/RUNG), consequence (LEARN/KNOW), incompatibility (RED/YELLOW/GREEN/...), complementarity (MALE/FEMALE), antonymy (LONG/SHORT), converseness (HUSBAND/WIFE), reversiveness (LEARN/FORGET) (cf., in addition to the works of Lyons cited above, Lehrer, 1974: 22–30 and the detailed treatment in Cruse, 1986). Several of these categories can be further divided into sub-types. Since paradigmatic relations are established between lexical items which occur in the same linguistic context, they hold between members of the same syntactic class, as illustrated above. In the case of comparable relations holding between members of different classes, we may speak of paradigmatic quasi-relations: thus, the adjective RED is a quasi-hyponym of the noun COLOUR (cf. *What colour is it? It's red*) and the adjective SWEET a quasi-hyponym of the verb TASTE (cf. *What does it taste like? It's sweet*) (cf. Lyons, 1977: 298–9).

The meaning of a lexical item thus includes a set of paradigmatic relations holding between it and other items in the same system. The size of this set will vary: in all languages, some areas of the vocabulary are more highly structured than others, and this will be reflected in a larger bundle of paradigmatic relations in such areas. Thus, for obvious cultural reasons, in most varieties of English terms for livestock are more highly structured semantically than, say, terms for sea creatures: COW is related paradigmatically to a variety of items such as BULL, BULLOCK, HEIFER, CALF, CATTLE, MOO-COW, UDDER, HERD, etc., whereas a term like OCTOPUS enters into a much smaller set of comparable relations. Some relations are intuitively more central than others: the relation of COW with BULL, for instance, seems more central than that with HEIFER, which is perhaps restricted to the passive vocabulary of many present-day English speakers. Clearly, the statement of paradigmatic relations may be more or less exhaustive, but a satisfactory statement will include all central relations.

2.1.2 Syntagmatic relations

Here we refer to the semantic relations which hold between a lexical item and other lexical items with which it can co-occur in syntactic constructions.[5]

A given lexical item is associated with a set of items with which it normally co-occurs in various syntactic constructions, and, in principle, the statement of the linguistic context of paradigmatic relations will involve the specification of the members of such a set. McIntosh has called this set the 'range' of a lexical item (1961: 330), and the relationship between a lexical item and its range is normally termed 'collocational'. Thus, in English, the adjective SWEET normally collocates, in different constructions, with nouns such as APPLE, TEA, TASTE and PERSON, and verbs such as TASTE and SMELL. Case relational terminology ('agent', 'patient', 'instrument', etc.) has been applied in the description of certain kinds of collocational relations (cf. Kittay and Lehrer, 1981; Kittay, 1987: 244–8).

Lexical items vary widely in their collocational range (cf. Lyons, 1977: 261). At one extreme, we find cases where one item necessarily collocates with another in a given construction, and in this case we may speak of a relation of syntagmatic implication. Thus, to cite an example from Lyons (1968: 422), we may say that BITE syntagmatically implies TEETH in *I bit him with my false teeth.* Porzig (1950: 68) has discussed such relations (under the heading of 'syntagmatic fields') in a well-known passage:

> Womit *beisst* man? Natürlich mit den *Zähnen.* Womit *leckt* man? Selbstverständlich mit der *Zunge.* Wer *bellt?* Der *Hund.* Was *fällt* man? *Bäume.* Was ist *blond?* Menschliches *Haar.* ('What does one *bite* with? With the *teeth,* of course. What does one *lick* with? Obviously, with the *tongue.* What *barks?* A *dog.* What does one *fell? Trees.* What is *blond?* Human *hair.*')

Porzig also mentions cases where the relation is less clear-cut: the verb REITEN 'ride' syntagmatically requires a noun belonging to the set of animal terms, but there are various possibilities within that range. In this connection, Coseriu has made a distinction between (syntagmatic) implication, where a single lexical item is involved, and (syntagmatic) selection, where an inclusive term ('archilexeme') and its subordinates are equally permissible (1971: 210). In fact, at least in English, cases of unique implication seem to be rare: thus, BLOND collocates with MOUSTACHE, BEARD, WIG as well as with HAIR;[6] FELL with ELM, ASH, etc. as well as with TREE; and BARK with FOX, BITCH, and ANIMAL, etc. as well as with DOG. It is clear, however, that such alternatives commonly stand in some kind of paradigmatic relation to each other: thus, a relation of partonymy exists between HAIR and MOUSTACHE, BEARD, WIG (cf. *the hair of his moustache*, etc.), and hyponyms and superordinates figure in most of the other examples. To return to the example of the false teeth, *I bit him with my false incisors* is a meaningful, if improbable, English utterance, and perhaps there is a general principle at work here to the effect that items in certain paradigmatic relations are always admissible, if present.

Relations of this kind clearly form an important part of the meaning of the lexical items to which they apply. Many items, of course, have much wider collocational freedom. There are limits, however, to the normal collocational range of all items, and these need to be indicated in a systematic semantic description. Where two lexical items may not normally collocate in a given construction, we may speak of syntagmatic incompatibility: thus, SWEET collocates with the verbs TASTE and SMELL *(This food tastes/smells sweet)* but not normally with FEEL. Many cases of syntagmatic incompatibility follow clearly from intralingual or extralingual meaning, involving tautology (**glasses and spectacles*, **I bit him with my teeth*, **I bought it for money*) or contradiction (**sweet, sour taste*, **I bit him with my false eyelashes*, **I bought it free*). While infringements of these incompatibilities may sometimes occur, they call for special interpretations and produce special effects, attesting to the presence of the constraints in normal circumstances. Other syntagmatic incompatibilities are more idiosyncratic, being fixed by usage rather than following predictably from meaning: thus *put a question* and *make a suggestion* are normal collocations in English, whereas **put a suggestion* and **make a question* are not (Allerton, 1984: 31–3). These idiosyncratic collocations are commonly language-specific, and lie in the grey area between free combinations and idioms.

2.2 Extralingual meaning

The previous section has been concerned purely with the intralingual semantic structure of the vocabulary, in terms of relations holding between lexical items.

But language is used in the world, and lexical items relate to aspects of this world: in particular, lexical items are applied to extralingual categories of entities, qualities, actions, events and states, and the relation between an item and such categories (commonly termed 'denotation')[7] is normally understood as constituting a central part of its meaning. Thus, we have seen in the previous section that the intralingual meaning of the adjective SWEET includes, for example, its incompatibility with SOUR in collocation with nouns such as APPLE, and so on. However, intralingual relations alone do not tell us how the item is normally applied to the world: what type of quality do English speakers normally use the word SWEET to describe? The intralingual structure of the lexical system indicates that this will be a distinct quality from that described by SOUR, but a satisfactory statement of the meaning of SWEET must clearly include a specification of its denotation.

Thus, while intralingual meaning is concerned with word-word relations, extralingual meaning involves word-world relations in that it brings into consideration the world of our experience.[8] Our discussion of how lexical items 'hook' on to this world will focus on two main points. Firstly, we stress that denotation cannot be usefully discussed in isolation from intralingual relations: intralingual and extralingual meaning are complementary. Secondly, we point to the study of focal exemplars and of the axiomatic propositions with which they are commonly associated as providing a valuable approach to the investigation of denotation in many extralingual domains, including taste.

With regard to the first question, the well-known futility of attempting to teach extralingual meaning purely by gestural methods (i.e. ostensive definition) reflects the necessity for a lexical item to be placed, implicitly or explicitly, in some kind of intralingual matrix before its denotational function can be properly dealt with. As a general principle, the denotation of a lexical item cannot usefully be discussed in isolation from its place in the structure of the vocabulary. To illustrate with a simple example, a treatment of the denotation of RED in terms of the psychophysical parameters of hue, brightness and saturation will produce different results according to whether we are dealing with the lexical item as belonging to a lexical system with ORANGE, GREEN, CRIMSON, etc. and collocating with items like CAR and DRESS, or with the lexical item as a member of a lexical system with BLOND, FAIR, GINGER, etc. and collocating with HAIR; for the colour quality of hair to which we can appropriately apply the term RED overlaps, psychophysically, with the range of colour qualities to which we apply the terms ORANGE and BROWN where ordinary objects are concerned.

Turning to the second point, the central fact to be accounted for in a theory of denotation is that the speakers of a language apply lexical items to various aspects of the world and are understood to have done so by other speakers: the conventions of a language-community include a set of denotational categories within the range of each of which a given lexical item is appropriately applied. How can the range of these categories be specified? How do speakers come to apply lexical items appropriately, given that the range of the categories involved is wider than any single instance of them?

We approach these questions by way of a focal model of denotation, in which the range of entities, qualities, actions, etc., to which a given lexical item can be appropriately applied is exemplified by a focal exemplar of the category concerned. Increasing attention has been paid to the importance of such denotational foci in recent work on semantics, originally by Berlin and Kay (1969), who demonstrated the high degree of agreement across speakers on the location of foci for colour categories in various languages. This work has subsequently been extended significantly, notably by the studies of Rosch and her colleagues on the internal structuring, in terms of prototypical and deviant membership, of a variety of perceptual and semantic categories (cf. e.g. Rosch, 1977). Foci clearly offer the semanticist a useful and important point of reference in the study of denotation.

In terms of this approach, let us consider a simple case of the appropriate application of a lexical item: the colour term RED (as in an utterance such as *What are those red flowers called?*). How can the relation between this term and its corresponding denotational category be handled in terms of focal exemplars? In this case (RED is here a word of relatively 'concrete' meaning), it is in principle possible to point to a typical instance of the colour category concerned or, perhaps, to produce one by pictorial means. (Needless to say, since ostensive definition is impossible, this method assumes that the learner of the category has the necessary cultural and linguistic competence, first, to grasp the point of such an exercise and, second, to understand the parameter of the indicated object that is involved, i.e. colour as opposed to shape, etc.) On the evidence of Berlin and Kay referred to earlier, different speakers of English would be likely to agree on the choice of typical example, and the learner would in this way gain a useful anchoring point for his application of the term RED. If, however, there is no focal exemplar to hand, or we do not have access to graphic materials, it is to be noted that a verbal alternative is available as a means of indicating the focus: namely, the utterance of the sentence *Blood is red* as a generic statement. Again assuming that he has the necessary degree of linguistic sophistication (and, of course, that he is familiar with the substance concerned), this method would serve to provide the learner with a focal exemplar for the category RED. Here too, we suggest, speakers of English would tend to agree on this choice of verbal example: *Blood is red* has, as it were, axiomatic status for speakers of English as a focus-indicating proposition. Similar axioms readily come to mind for certain other colour terms (*Snow is white*, *The sky is blue*, *Grass is green*), and are indeed often drawn upon in dictionary definitions (e.g. BLUE: 'the colour of the unclouded sky', WHITE: 'the colour of pure snow', etc.). (Interestingly, other colour terms, especially those classified as later additions in Berlin and Kay's evolutionary hierarchy (cf. 1969: 22–3), are typically defined in terms of earlier terms (PINK: 'a whitish red', BROWN: 'a dark reddish colour', GREY: 'a mixture of black and white', etc.); similar remarks also apply to non-basic terms such as AQUA ('greenish-blue'), etc.)

Such extralingual semantic axioms are of obvious importance to the study of denotation, both in 'concrete' areas of the vocabulary such as the perceptual domains discussed here and, *a fortiori*, in the case of 'abstract' domains where more direct methods of exemplification are less practicable. Aschmann's study of Totonac smell categories, as reported by Sturtevant (cf. 1.2), appears to make use of a focal approach, listing typical exemplars of the categories concerned, and Sturtevant's comments leave the impression that, for him, this ranks a poor second to scientifically-based analyses of domains (1964: 119). However, even within human sense perception, there are few domains where scientific understanding is sufficiently advanced to provide a framework for the discussion of denotation: as we have seen, many aspects of taste have so far proved highly resistant to scientific analysis. Moreover, even where such studies are helpful, they may have little relevance in themselves to average native speakers in their application of lexical items; speakers acquire and use the rules of denotation of their native language unaided by such scientific analyses. By contrast, it seems clear that focal exemplars, and their typical verbal formulations, do have a role to play in this regard, by readily providing initial 'givens' for denotation and indicating the points at which lexical items 'hook' on to the surrounding world most securely.[9]

For this reason, axioms constitute an aspect of semantic knowledge that deserves to be recorded as part of the data in investigations of denotation. Needless to say, interesting cross-linguistic differences in such axioms can be expected: even with domains that are assumed to be universally encoded, differences in natural and cultural emphasis will doubtless be reflected in the occurrence within axioms of different focal exemplars; where more culture-bound categories are concerned, clearly the possibilities are legion.

We still need to consider the problem of how speakers of a language learn the range of a category: a focal exemplar, by its very nature, can only exemplify, it cannot delimit. Thus, armed merely with a knowledge of the focal exemplar of RED and an instruction such as 'Apply RED to instances identical with, or similar to, the focal exemplar', a learner would never arrive at appropriate application of the term since he has no way of telling at what point a given instance ceases to be similar. Denotational categories are not learned in isolation, however, but within the intralingual framework of lexical systems: as soon as we have more than one focus in the colour domain, a basis exists for judging intermediate cases in terms of similarity and for determining the outer boundaries of each category (cf. Harrison, 1973: 66–9; Kay and McDaniel, 1978: 623–4).

In summary, we propose to make use of focal exemplars and associated semantic axioms in our treatment of the denotation of Japanese taste terms. Not only does such an approach seem the only practical one at the present stage of understanding of human taste perception; insofar as focal exemplars have relevance to the way that the speakers of a language come to apply lexical items appropriately to aspects of the world around them, they clearly constitute a valid and important subject for semantic investigation in their own right.

2.3 Basic and extended meanings

In this section we consider briefly how the question of semantic extension may be approached in the present framework.

The distinction commonly made between basic and extended meanings is in general an intuitively satisfying one and, as such, no doubt has a sound basis in linguistic fact. The primary task of a semantic theory in this area must be to examine the facts in a wide variety of cases and to explicate and refine the distinction in the light of the findings. In terms of the present framework, this entails making detailed studies of the intralingual and extralingual relations which hold in various cases. One result of such studies would be a typology of semantic extensions.

Semantic extension involves a change in meaning whereby items that are 'basically' applied in one domain are further applied in other, often more abstract, domains. An example is the extension, illustrated earlier, of the English adjective SWEET from the domain of taste to the domain of disposition or behaviour. Such extensions raise interesting and familiar questions for semantic research: Are any patterns of extension universal? Are any universally absent? A study by Williams (1976) has considered such questions for cases of synaesthetic extension, where adjectives basically applied in the domain of one of the human senses are transferred to others, and has identified definite tendencies across languages (including Japanese) within this area. In the case of taste terms, his findings are that, in general, they may be extended to the domains of smell or sound (and, via sound, to colour); in the reverse direction, only touch terms may be extended from other senses to the domain of taste. These findings have also been discussed by Lehrer (1978), who examines semantic extension in English adjectives which are applied to wine.

It frequently happens that paradigmatic semantic relations which hold in the basic domain are carried over, to a greater or lesser degree, into the extended domain. Indeed the fact that this is so reflects the importance of recognizing the intralingual organization of the vocabulary in the form of semantic systems (cf. Lehrer, 1974: 110–19; 1978). In this respect, attention needs to be directed to determining what types of modification, both quantitative and qualitative, of basic intralingual relations occur in various cases. Where some, but not all, members of a system are extended to a new domain, does this perhaps indicate, as Lehrer suggests (1974: 115), that the recalcitrant terms are less central members of the basic system? We shall give consideration to such questions in examining extended meanings of Japanese taste terms in the course of the semantic analysis.

2.4 Stylistic factors

Many of the lexical items in a language may be characterized in terms of stylistic properties of various kinds. These relate to aspects of the situations in which lexical items are appropriately used, and involve such matters as the social relationships of the participants in the interaction, and the medium and genre of communication employed. The analysis of evaluative taste terms in Japanese in Chapter 4 will necessitate consideration of stylistic factors, relating primarily to the sex of participants and the degree of social distance between them. In this section we wish to consider some general aspects of the interaction of meaning and style, and also to outline the relevant facts of stylistic variation in Japanese.

2.4.1 Meaning and style

The incidence of lexical stylistic differentiation varies, both across languages and across particular areas of the vocabulary. As a simple example of sporadic differentiation of this kind, let us consider the case of (British) English CIGARETTE and FAG. Here FAG is clearly stylistically marked, as informal, but it stands as an isolated example of such marking in this particular domain. In meaning, CIGARETTE and FAG are apparently identical: both enter into the same intralingual relations (incompatibility with CIGAR, etc.) and have the same denotation. Whether we treat them as synonyms will depend upon whether we choose to set up paradigmatic relations relative to particular situational, as well as linguistic, contexts (cf. 2.1.1); in either case, a comprehensive description of the behaviour of lexical items will certainly need to take account of stylistic properties. It seems that synonymy within a single style is rare as a lexical relation: GLASSES and SPECTACLES, for example, suggested as synonyms in 2.1.1, are also stylistically differentiated in terms of formality. Allan (1986: 178–9) proposes a relation of 'cross-varietal synonymy' for contextually differentiated cases of this kind, and Cruse (1986: 88, 265–91) speaks of 'cognitive synonymy' (as opposed to 'absolute synonymy') where truth-conditions are maintained despite differences such as those under discussion here. Clearly, these notions could be extended to other types of paradigmatic relation (cf. Cruse, 1986: 290–1).

Other cases of stylistic differentiation may be much more pervasive. Here, there will be clear differences in both paradigmatic and syntagmatic relations between items from different styles. An extreme example of such systematic differentiation is provided by the well-known case of the mother-in-law language reported by Dixon (1971) for the Dyirbal language of North Queensland. When conversing within earshot of a certain class of relatives, a speaker of that language is obliged to use a special 'mother-in-law' language in place of everyday Dyirbal. The relationship between the two varieties is as follows: while phonology and grammar are identical in both, all lexical items in everyday Dyirbal are replaced by a distinct set of lexical items in the mother-in-law language. The fact that the latter set is far smaller in number means that the relationship of lexical items between the two varieties is largely many-to-one: in general, several distinct lexical items in everyday Dyirbal are replaced by one item in the mother-in-law language, although cases of one-to-one relationship also occur. Dixon uses this interesting situation to throw light upon the semantic structure of everyday Dyirbal by examining the relationships of 'synonymy' (one-to-one correspondence) and 'hyponymy' (many-to-one correspondence) that hold between lexical items in the two varieties. In terms of the present discussion, these relations are cross-varietal or cognitive, and one would expect additionally to find absolute hyponyms (and perhaps synonyms) within everyday Dyirbal – if not, given its special lexical paucity, within the mother-in-law language. Syntagmatically, lexical items from these two styles will differ systematically in collocational range: in general, stylistic congruity is a further determining factor on normal collocability (cf. 2.1.2) although, as Cruse points out (1986: 100–1), its constraints (in cases of more sporadic variation) may not necessarily be statable within the confines of syntactic constructions.

Though the extent of stylistic lexical differentiation in Dyirbal is clearly unusual, many other languages exhibit stylistic elaboration within particular areas of the vocabulary. Consider a lexical item such as English GEE-GEE, belonging to the stylistic variety known as baby talk. The variation here is more systematic than in the case of CIGARETTE and FAG: in this variety, ordinary lexical items from various domains, including familiar farmyard and domestic animals, are regularly replaced by stylistically marked terms. And this has important semantic consequences: for while GEE-GEE and HORSE have the same denotation, they quite clearly have different intralingual relations. HORSE is a hyponym of ANIMAL, itself has hyponyms such as STALLION, and is incompatible with items such as COW. GEE-GEE, on the other hand, does not form part of a lexical hierarchy, and is incompatible with such items as MOO-COW, BOW-WOW and the like; if GEE-GEE can be said to have an absolute synonym, presumably it is HORSIE rather than HORSE. While this may appear to be a somewhat trivial example, it highlights differences in intralingual structure between varieties which are easily overlooked. It also serves to illustrate the point that, not only different languages, but different areas of the vocabulary within the same language, may present different problems of interaction between meaning and style.

2.4.2 Japanese honorifics

As is well known, Japanese is characterized by an intricate system of semantic-stylistic distinctions generally known as 'honorifics' (*keigo*). While the reflexes of this system are primarily grammatical, it also has lexical repercussions of relevance to our study, and it is appropriate to give a brief outline of the system here.[10]

The Japanese system can be viewed in terms of two intersecting axes: the axis of reference, and the axis of address. On the axis of reference, a 'neutral' level contrasts with an 'honorific' level, where honorific forms are used to indicate respect by the speaker towards a person spoken about (who may or may not be identical with the addressee): broadly, subject-honorific forms are used with reference to the respected person or to actions, entities, etc. where the respected person is the subject or possessor, and object-honorific forms are used with reference to the speaker's actions affecting the respected person. Neutral forms are, as expected, unmarked with regard to these distinctions and are used where no special respect is being shown. On the axis of address are distinguished two basic speech styles, 'informal' and 'formal', the choice between these being determined mainly by the degree of social distance between the speech participants: in general, the informal style is used between familiars, the formal style elsewhere.

The formal distinctions involved are realized most extensively in verbs. The typical situation can be illustrated with the verb YOMU 'read', as follows:

(1)

		Axis of address	
		Informal	Formal
Axis of reference	Neutral	*yomu*	*yomimasu*
	Subject-honorific	*oyomi ni naru*	*oyomi ni narimasu*
	Object-honorific	*oyomi suru*	*oyomi shimasu*[11]

In certain cases, however, instead of the regular morphological changes illustrated, suppletion occurs. The corresponding table for the verb IU 'say', for example, has the following forms:

(2)

		Axis of address	
		Informal	Formal
Axis of reference	Neutral	*iu*	*iimasu*
	Subject-honorific	*ossharu*	*osshaimasu*
	Object-honorific	*mooshiageru*	*mooshiagemasu*

Here the expected honorific (informal) forms **oii ni naru* and **oii suru* do not occur and are replaced by the forms *ossharu* and *mooshiageru*. The suppletion involved here is one-to-one (i.e. *ossharu* and *mooshiageru* occur only in relation to the verb IU), but many-to-one relationships are also found: neutral *iku* 'go', *kuru* 'come' and *iru* 'be located (of animate entities)', for example, are all suppleted by *irassharu* or *oide ni naru* at the subject-honorific level.

The axis of address, less studied in this respect, is sometimes associated with true lexical variation, as opposed to suppletion against a background system of regular morphological changes. In the first place, a small number of common verbs have extra-formal variants, used to signal high formality of the speech situation: examples include *gozaimasu* (informal *aru* 'exist, be located (of inanimate entities)'), *orimasu* (*iru* 'be located (of animate entities)') and *mairimasu* (*iku* 'go' and *kuru* 'come'). Secondly, and more centrally to the present study, certain lexical items have (in the case of inflected items, in addition to regular informal and formal forms) stylistic lexical variants which we shall classify as 'masculine'. Such variants are almost exclusively used by male speakers, in highly informal situations, whereas the ordinary, unmarked forms are used by speakers of both sexes alike. Several basic lexical items from domains relating to food show variation here, including the verb TABERU 'eat', which has a masculine variant KUU, and the noun GOHAN 'boiled rice, meal', which has the variant MESHI. The phrase ONAKA GA SUKU 'feel hungry' has a masculine equivalent in HARA GA HERU, and as we shall see, the evaluative adjective OISHII 'good-tasting' has a masculine variant in UMAI. It is thus possible to envisage male informal utterances containing the lexical items KUU, MESHI, HARA GA HERU and UMAI being reported by a female speaker (or by a male speaker in more formal circumstances) using TABERU, GOHAN, ONAKA GA SUKU and OISHII, resulting, within a restricted area of the vocabulary, in a miniature parallel of the Dyirbal situation. We should add here that 'soft' (or 'feminine')[12] lexical variants are also found in Japanese, used predominantly, though by no means exclusively, by women and typically formed by the prefixation of O- (thus, YASAI 'vegetables'/OYASAI, etc.). Although these will not concern us centrally in this study, it is worth noting that the domains of food and eating are also prominently represented here (cf. Shibata, 1957: 46).

The stylistic behaviour of the evaluative taste terms UMAI and OISHII will be examined in greater detail in 4.5.

3 *Methodology*

In 3.1 we outline the methodological approach adopted in the subsequent delimitation and semantic analysis of Japanese taste terms. Details of the collection of data are given in 3.2, and the representational conventions employed for Japanese in the study are summarized in 3.3.

3.1 Delimitation and semantic analysis

The theory of lexical meaning presented in Chapter 2 is structuralist in nature: the meaning of a lexical item in a given language is basically a function of the intralingual semantic relationships which it contracts with other items. From this it follows that the elucidation of lexical meaning must necessarily be approached from within the language concerned. Such a view is supported by the well-known fact that lexical systems encoding 'corresponding' domains in different languages commonly prove to be anisomorphic, both in the placement of the boundaries of the domain and in its subdivision.[1]

Before undertaking this study, the investigator was familiar, to a greater or lesser degree, with a range of Japanese lexical items whose meaning seemed to relate in some way to 'taste' in the widest sense, such as UMAI 'good-tasting', AMAI 'sweet', SHIBUI 'astringent', KOOBASHII 'fragrant', ABURAKKOI 'fatty', MIZUPPOI 'watery, insipid', SAPPARI-SHITE IRU 'clean-tasting', KORIKORI-SHITE IRU 'resilient', SUJIPPOI 'stringy', etc. What was not known, and what the investigation was intended to discover, was how such terms were semantically organized within the Japanese vocabulary. Do all such terms belong together in a single lexical system, or not? How is this to be shown? How can we decide in a linguistically principled manner what counts as a taste term in Japanese?

In this study, we approach these problems by means of a methodology based upon question-and-answer procedures.[2] Underlying their use is the notion of a question as a semantic range-finder: in general, a given question asks for information within a restricted semantic range, and only information within that range is relevant in answer to it. Examining the lexical aspects of appropriate questions and responses relating to a given domain thus provides a valuable approach to understanding its semantic structure.

Once the desirability of a question-and-answer approach to delimitation has been recognized, ascertaining the most appropriate question for the particular domain being investigated is an important task in itself. In the case of this study, we must discover how Japanese ask questions about taste: for example, if we assume that Japanese has a question roughly equivalent to English *What does it taste like?*, then lexical items elicited in responses to that question can justifiably be viewed as constituting the lexical field of 'taste' in Japanese by virtue of their common occurrence. There is no guarantee, of course, that Japanese will have such a question at this particular level of generality, and this possibility merits methodological consideration. Since colour is a more familiar area of discussion, let us examine an imaginary case from that domain.

Let us consider the possibility of a language, whose colour terms we wish to delimit and analyse, that has no term corresponding to English COLOUR, and therefore no question equivalent to English *What colour is it?*, but only a question at a more general level such as *What does it look like?*[3] The first step here would be to determine what lexical items occurred in responses to that question: let us assume, naively, that they comprise lexical items corresponding to English GLOSSY, DULL, BLACK, WHITE, and RED. By virtue of their common occurrence, these items could be grouped together in a lexical field which, given the associated question, we might label 'external appearance'. It may be that further semantic subdivision within the field is not evident: in that case we should say that the language does not contain a lexical system of colour terms as such, but a field of external appearance terms which, denotationally, ranges over the psychophysical properties of hue, brightness, saturation and lustre. On the other hand, the use of supplementary question-and-answer procedures of a more refined type may reveal the possibility of subdivision into tighter lexical systems within the overall field. The approach here will be to use the lexical items already elicited and to ask questions of the type *Is X glossy?*, *Is X dull?*, and so on for the five lexical items. Examination of the response patterns may show, for example, that DULL and GLOSSY occur only as answers to questions containing each other; that is, we may find the question-and-answer pairs *Is X glossy? No, it's dull* and *Is X dull? No, it's glossy*, but not such pairs as *Is X glossy? *No, it's red*, etc. On this basis, DULL and GLOSSY fall together within the larger field in a lexical system, which we might label 'glossiness'. Assuming that the remaining terms, BLACK, WHITE and RED, fall into a similar group, these too would be seen as constituting a lexical system which, in view of the denotational properties concerned, could justifiably be labelled 'colour'. In such a case, then, a general question frame is available only at a relatively high level of the semantic hierarchy, but further sub-groupings are revealed by the use of supplementary question-and-answer procedures. Procedures of this type prove to be of particular importance in the analysis of descriptive taste terms in this study, and they will be employed in delimiting and subdividing the lexical field of taste in Japanese in Chapters 5 and 6.

Such procedures, besides having the theoretical virtue of attacking the vocabulary of the language from within, also have the methodological advantages of being a relatively straightforward method to use with informants, and of permitting the results obtained to be tested by other investigators. They appear to have application over a wide area of the vocabulary, and indeed to provide a valuable general approach in the investigation of lexical fields.

3.2 Collection of data

The variety of Japanese on which this study is based is the modern spoken language of educated inhabitants of Tokyo. In particular, for practical reasons it is based chiefly on the speech of one informant, the wife of the investigator, who was raised and educated in Tokyo. The sex of the informant and her relationship to the investigator are reflected in certain stylistic features of the primary data: the informal style of speech is used throughout and, since the informant is female, masculine stylistic variants are not elicited (cf. 2.4.2).

As previously noted (cf. 1.3), we will expect the linguistic reflexes of the individual variation commonly recognized to exist in the area of taste perception to relate predominantly to the application of lexical items, especially evaluative terms, to particular food substances. As far as the intralingual relationships among taste terms are concerned, there seems to be no reason to suppose that this lexical field should differ from others: that is to say, one may expect that some variation will occur in respect of peripheral members of the field (with, for example, some individuals or groups possessing a more elaborate taste vocabulary than others), but that the structure of the core terms will be stable across speakers of the variety. One approach to investigating the degree and type of linguistic variation that does occur would clearly be to carry out additional studies of the present kind for different idiolects. As one potentially relevant factor here, let us recall the geographical variation in taste terms recorded by the Linguistic Atlas of Japan (cf. 1.2): we shall see that in at least some cases this gives rise to stylistic variation within idiolects, and this may constitute one area of instability across speakers.

The primary data of the study were elicited, using question-and-answer procedures of the type outlined above, over a period of several months. In general, questions were put to the informant on several occasions in order to determine the stability of responses, and only question-and-answer pairs repeatedly checked in this way were admitted as data and appear in the study.[4] Denotational axioms and taste norms were established, relatively informally, by asking the informant to name typical food substances exemplifying given taste qualities (for axioms) or expected taste qualities for given food substances (for taste norms); here too, questions were put on several occasions, and only cases which showed a high degree of salience on different occasions are included as data. Sources of secondary data include observed conversations involving the informant, together with additional spoken and written language data from (a) a round-table gastronomic discussion published in the monthly women's magazine *Fujin Kooron*, (b) two popular science paperback publications dealing with cultural aspects of diet (Koono, 1974; 1977), and (c) the published proceedings of a conference on culture and diet including contributions by linguists, anthropologists and food scientists (Shibata and Ishige, 1983). Treatments of Japanese taste terms in linguistic publications mentioned in 1.2 were also consulted, as were Japanese dictionaries.

As far as the membership of the lexical field is concerned, the appropriateness of the lexical items elicited is underwritten by the question-and-answer methodology adopted: that is to say, no lexical items appear in the analysis which are not 'true' taste terms in a linguistically definable sense. The problem of exhaustiveness remains: is the list of elicited terms 'complete'? No methodology can guarantee exhaustiveness; over the period during which this study was carried out, however, no additional lexical items of Tokyo spoken Japanese belonging to the field as delimited were discovered, either in observed conversations or in secondary sources of data. Putative exceptions primarily involve written language items or items from neighbouring fields, and these will be mentioned at appropriate points in the study.

3.3 Conventions of representation

Throughout this study, Japanese is rendered orthographically in the Hepburn system of romanization, with one modification: for typographical convenience, long vowels are represented as a succession of two short vowels (*aa*, *ii*, *uu*, *ee*, *oo*) rather than by means of macron superscripts placed over single vowels.

Japanese data is given in normal type, with taste terms under consideration italicized for ease of recognition (Sono wain wa *oishii*). English glosses are enclosed in single quotes ('That wine tastes good'). As well as being provided with a relatively natural English gloss, primary data is generally given an interlinear rendering as in the following example:

Meron wa donna aji ga suru?
MELON/isolator [5]/WHAT KIND OF/TASTE/subject/BE PRESENT, non-past
'What kind of a taste do melons have?'

In such renderings, slashes correspond to the separation of forms in the original Japanese: in general, lexical information is given in small capitals, grammatical information in lower case; where more than one type of information is involved, as in the case of the final form of the example, the items of information are separated by commas. Interlinear renderings are abbreviated or omitted in clear cases, such as in series of data of identical grammatical structure.

4 *Evaluative taste terms in Japanese*

In this chapter we treat the three lexical items OISHII 'good-tasting', UMAI 'good-tasting' and MAZUI 'bad-tasting', which together constitute a set of purely evaluative taste terms in Japanese. [1] In 4.1 we discuss the delimitation of these terms. Their grammatical characteristics are surveyed in 4.2, and their syntagmatic and paradigmatic meaning relations are described in 4.3 and 4.4, respectively. These items are characterized by stylistic differentiation of a particularly complex kind, and the factors involved are examined in 4.5. 4.6 addresses the question of their extralingual meaning, and extended meanings of the terms are considered in 4.7.

4.1 Delimitation

The terms OISHII, UMAI and MAZUI regularly occur in responses to the question-frame *X wa oishii/umai?* (X/isolator/GOOD-TASTING, non-past) 'Is/are X good-tasting?'.[2] The use of such a frame is a typical method of asking for evaluative judgements of the palatability of food substances: X in the frame is occupied by nouns denoting food substances or, alternatively, eating-places. [3] The following are typical examples of questions and responses:

(1) Kono momo wa *umai*? Ee, suimi ga atte *oishii*.
THIS/PEACH/isolator/GOOD-TASTING, non-past/YES/TARTNESS/subject/BE PRESENT, conjunctive/GOOD-TASTING, non-past
'Are these peaches good?' 'Yes, they have a tartness and taste good.'

(2) Asuko no tempura wa *umai*? Iyaa, ammari *oishiku* nai.
THAT PLACE/genitive/TEMPURA/isolator/GOOD-TASTING, non-past/NO/(NOT) VERY/GOOD-TASTING/negative, non-past
'Is the *tempura* there good?' 'No, not very.'

(3) Sono wain wa *oishii*? Un, *umai*.
THAT/WINE/isolator/GOOD-TASTING, non-past/YES/GOOD-TASTING, non-past
'Is that wine good?' 'Yes, it is.'

(4) Asuko wa *umai*? Iya, *mazui* yo.
THAT PLACE/isolator/GOOD-TASTING, non-past/NO/BAD-TASTING, non-past/emphasizer
'Is the food at that place good?' 'No, it isn't.'

As shown by (1), descriptive taste terms may occur in fuller responses to the frame, but in such cases they are typically accompanied by one of the evaluative taste terms: thus, in (1), the conjunctive form of the expression SUIMI GA ARU 'has a tartness' is followed by a form of OISHII. The evaluative taste terms, however, do not occur in direct replies to the question-frame *X wa donna aji ga suru?* 'What kind of a taste does X have?', used in the delimitation of descriptive taste terms in Chapter 5.

It is worth noting here that, in these items, Japanese possesses a basic set of taste evaluators distinct from the more general evaluators, notably II 'good'. In this it differs, for example, from English, which tends to use general evaluative items here such as GOOD, NICE, SUPERB, AWFUL, etc. One consequence of this is that an English sentence such as *These eggs are good* has, out of context, more than one Japanese equivalent: *Kono tamago wa oishii/umai* 'These eggs are good(-tasting)' with OISHII/UMAI and *Kono tamago wa ii* 'These eggs are good (in quality, size, etc.)' with II. (Where necessary, of course, English has syntagmatic recourse to the more explicit *These eggs taste good.*) More specific English items such as TASTY, PALATABLE, YUMMY, SCRUMPTIOUS, etc. are restricted in various ways vis-a-vis OISHII or UMAI. This also applies to DELICIOUS, which is frequently misused as a general translation-equivalent for these terms. Semantically, DELICIOUS is an implicit superlative (Cruse, 1986: 216–17), standing at the end-point of a scale and relating to NICE, etc., as ENORMOUS relates to BIG; as such it may not be modified by graders (VERY, etc.) unlike OISHII and UMAI which, as we shall see, are gradable terms.

4.2 Grammatical properties

Here we note the main morphological and syntactic characteristics of OISHII, UMAI and MAZUI.

Like the majority of items treated in this study, these three terms are *i*-adjectives.[4] The principal inflected forms of these adjectives may be illustrated for OISHII as follows:

(5)

Non-past	*oishii*
Past	*oishikatta*
Adverbial	*oishiku*
Adverbial (extra-formal)	*oishuu*
Conjunctive	*oishikute*

The stems *oishi-*, *uma-* and *mazu-* are synchronically all morphologically simple. Non-past and past forms of *i*-adjectives may function alone as predicators (*Kono tamago wa oishii/oishikatta* 'These eggs are/were good-tasting'); in attributive function, they precede their head noun (*oishii tamago* 'good-tasting eggs').

There is an important grammatical subdivision in Japanese between subjective and objective adjectives. The most obvious syntactic characteristic of subjective adjectives is that in declarative sentences they may be predicated directly only of first-person experiencers: taking SAMUI 'cold (as a bodily sensation)' as an example, *Boku wa samui* 'I am cold' is grammatical, but **Kimi wa samui* 'You are cold' and **Ano hito wa samui* 'He/she is cold' are not. In their place, recourse must be had to devices such as presumptive or evidential constructions (*Kimi/Ano hito wa samui-daroo/samusoo da* 'You/he/she must be cold/look(s) cold'), reported speech (*Ano hito wa samui to itte iru* 'He/she says that he/she is cold') or the use of a derived verb formed by the suffixation of -GARU (*Ano hito wa samugatte iru* 'He/she is showing signs of feeling cold'). This last derivational process in general provides a useful morphological indicator of the sub-class.[5] Semantically, subjective adjectives denote psychological states and reactions: observers do not have direct access to such states as experienced by others, and in these cases the grammar of Japanese, unlike that of English, requires that reference be made explicitly to their external manifestations rather than to the states themselves.

The subdivision thus involves a cluster of semantic, syntactic and morphological criteria comparable to those adduced for English by Dixon (1982) in his cross-linguistic survey of adjectives discussed in 1.3. Its relevance here is that both evaluative and, to a lesser degree, descriptive taste adjectives in Japanese share the characteristics of the subjective sub-class. In this they are clearly differentiated from colour adjectives, and the distinction may be seen as a reflection in linguistic structure of the greater subjectivity of taste perception suggested earlier (cf. 1.3). Though not echoed in English, this division, particularly as it involves descriptive taste terms, thus provides additional linguistic support for the general separation of colour and taste adjectives in the languages of the world.

Subjective features of evaluative taste terms will be discussed further in 4.6, and corresponding features of descriptive taste terms in the relevant sections of Chapter 5. In this section we may note with respect to derivational morphology that all three evaluative adjectives produce derived verbs in -GARU (OISHIGARU/ UMAGARU 'show signs of finding good-tasting'; MAZUGARU 'show signs of finding bad-tasting').[6] Japanese adjectives also regularly derive abstract nouns ending in the suffix -SA, and some adjectives additionally derive nouns (described by Martin (1975: 911) as 'somewhat less abstract') in -MI. The present items have associated nouns as follows:

(6)

OISHII:	OISHISA
UMAI:	UMASA, UMAMI
MAZUI:	MAZUSA

UMAI thus has nouns of both forms. Finally, like most adjectives the present terms derive evidential adjectives in -SOO-NA: OISHISOO-NA 'looking or seeming good-tasting', etc. Semantic and stylistic aspects of these derivatives will be discussed further below.

4.3 Syntagmatic relations

In general, as indicated above, the three items commonly collocate in subject-predicator or modifier-head constructions with nouns denoting food substances and eating-places. In addition to items denoting food substances in the strict sense, they collocate with items denoting other substances which are commonly ingested, notably nouns denoting smoking articles [7] (e.g. TABAKO 'cigarette', HAMAKI 'cigar') and, in the case of OISHII and (questionably) UMAI, but not MAZUI, with the noun KUUKI 'air' (*Kuuki ga oishii na* 'Isn't the air good (to breathe)!'): physiologically, breathing is an ingestive process comparable to eating and drinking. All three terms collocate with lexical items denoting eating-places (*Ano mise wa mazui* 'That place is bad (with respect to the palatability of its food)').

As regards the limits of collocability, the set is syntagmatically incompatible with the noun NIOI 'smell': **oishii nioi* 'good-tasting smell', etc. In this they contrast with English terms such as DELICIOUS, which are readily applied to smells. In Japanese, NIOI collocates with the general evaluator II 'good' and items such as IYA-NA 'unpleasant'; in addition, if reference is specifically to the appetizing or unappetizing properties of a food smell, evidential derivatives of the present terms are used: *oishisoo-na nioi* 'an appetizing smell, i.e. a smell (of food) which smells as if the food is good-tasting', etc. These restrictions reflect the fact that, in extralingual applicability, these terms are specific to the domain of taste.

Other restrictions on the collocability of individual items result from stylistic factors mentioned in 2.4.2 and discussed further below. Thus, within the field of food substances already referred to, we find cases where OISHII and UMAI show different collocational ranges: an example is provided by the nouns GOHAN and MESHI (both 'boiled rice, meal'), where the patterns are as follows:

(7) GOHAN ———— OISHII/UMAI
MESHI ———— *OISHII/UMAI

As noted above, OISHII, rather than UMAI, sits more comfortably with the noun KUUKI 'air': this also appears to depend on stylistic factors and is discussed further in 4.5.

Of the derived forms, -GARU verbs (OISHIGARU 'show signs of finding good-tasting', etc.) collocate in object-predicator constructions with nouns denoting food and other ingested substances, but nouns denoting eating places are less natural here. The stylistic properties of OISHII and UMAI are shared by their derivatives, with the exception of UMAMI; this will also be taken up further in the discussion of stylistic factors in 4.5.

4.4 Paradigmatic relations

The items OISHII, UMAI and MAZUI are interrelated in terms of synonymy (OISHII/UMAI) and antonymy (OISHII/MAZUI and UMAI/MAZUI).

Synonymy may be defined in terms of bilateral implication: if two sentences of identical syntactic structure each imply the other, and differ only in that one has lexical item X where the other has Y, then X and Y are synonyms (cf. Lyons, 1968: 450). The following sentences meet the required structural conditions, differing only in the lexical choice of OISHII or UMAI:

(8) a. Kono ringo wa *oishii.*
THIS/APPLE/isolator/GOOD-TASTING, non-past
'This apple tastes good.'
b. Kono ringo wa *umai.*
'This apple tastes good.'

The sentences exhibit bilateral implication: the assertion of one precludes the denial of the other, and sentences such as (9a) and (9b) are adjudged as contradictory:

(9) a. *Kono ringo wa *oishii* kedo *umaku* nai.
THIS/APPLE/isolator/GOOD-TASTING, non-past/BUT/GOOD-TASTING/negative, non-past
'This apple tastes good but doesn't taste good.'
b. *Kono ringo wa *umai* kedo *oishiku* nai.
'This apple tastes good but doesn't taste good.'

(9a) and (9b) are also stylistically incongruous, since OISHII and UMAI are stylistically differentiated. Their synonymy is thus cross-varietal or cognitive (cf. 2.4.1), and a fuller description of their usage must involve consideration of social factors to be discussed in 4.5.

Antonymy is the relation which holds between pairs of gradable opposites such as English BIG/SMALL and GOOD/BAD. Importantly, antonyms fall into different sub-types, and we preface discussion of the Japanese items with some general remarks on this topic. Thus, Lyons' basic test for antonymy involves the relation of bilateral implication between comparative sentences illustrated by the pair *Our house is bigger than yours used to be* and *Your house used to be smaller than ours is* (1968: 464): these sentences each imply the other, and BIG and SMALL are thereby shown to function as antonyms. However, not all pairs of items commonly considered as antonyms behave in the same way in relation to this test. If we compare the pair GOOD and BAD with the above pair BIG and SMALL, the following examples show that there is a lack of parallelism between them:

(10)a. My car is bigger than yours.
b. Your car is smaller than mine.

(11) Your car is smaller than mine, but they're both big cars.

(12)a. My car is better than yours.
b. Your car is worse than mine.

(13) *Your car is worse than mine, but they're both good cars.

With BIG/SMALL, (10a) and (10b) each imply the other, and furthermore (11) is an acceptable sentence. With GOOD/BAD, on the other hand, although (12b) implies (12a), (12a) may not necessarily imply (12b), notably in the case where it is understood that both are good cars. This is shown clearly in the unnaturalness of (13), where that understanding is made explicit; instead of (13), the natural sentence is exemplified in *Your car is not as good as mine, but they're both good cars*. (12a) and (12b) each imply the other if there is an understanding that both are bad cars. Unlike BIG/SMALL, then, GOOD/BAD do not meet Lyons' test in all contexts.

The issue is illuminated by Sapir in his article on grading (Sapir, 1944; references to Mandelbaum, 1949). There, Sapir distinguishes between two types of logical grading: grading with reference to a norm (type I), and grading with reference to terms of comparison (type II). He explains that: 'In type I...any "worse" or "bad" is worse than any "better" or "good"...But in type II..."worse" and "bad" do not need to be bad but may actually be good' (Mandelbaum, 1949: 126). On the basis of the above examples, *good*, *bad* and *worse* thus appear to belong to type I, *big*, *small*, *bigger*, *smaller* and *better* to type II. Sapir continues:

> A warning: These are logical terms, not terms of actual usage, which exhibit great confusion...'good' follows type I, but 'better' follows type II, being equivalent to 'relatively better, not so bad', e.g. 'My pen is better than yours, but I confess that both are bad'...Interestingly enough, the correlatives of these terms do not exactly correspond...'Bad' and 'less bad', differing in this respect from 'good' and 'better', both follow type I; 'less bad' is still 'bad' but 'better' (with reference to another term) may be even worse... On the whole, usage tends to assign comparative terms to type II of grading, positive terms to type I of grading, though this never hardens into a definite rule. (Mandelbaum, 1949: 126–7)

Sapir's account may be re-cast as follows: Some positive terms are inherently graded with respect to an absolute norm (type I: e.g. *good/bad*), while others are inherently graded with respect to a relative norm (type II: e.g. *big/small*). As for comparative terms, some comparatives of type I positives imply their positive terms (e.g. *worse*: *X is worse than Y* implies *X is bad*), while some do not (e.g. *better*: *X is better than Y* does not imply *X is good*); comparatives of type II terms do not imply their positives. This provides a ready explanation for the unnaturalness of (13), which we quote again here:

(13) *Your car is worse than mine, but they're both good cars.

Your car is worse than mine implies *Your car is bad*, which in turn implies *Your car is not good*; this latter implication is contradicted in the second part of the sentence. In terms of this account, Lyons' test works without exception for pairs of comparative terms where neither implies its positive term: pairs of type II positives thereby qualify as fully-fledged antonyms. Where one of the comparative terms does imply its positive, the test works in one direction everywhere, but not in all contexts in the other: the lexical pairs concerned (e.g. GOOD/BAD) are thereby marked off as a sub-type of antonyms vis-a-vis central cases.

Sub-types of antonymy have been further examined by Cruse (1976; 1986). Noting the differences discussed above in the semantic relation between comparative and positive forms, Cruse (1986: 206–14) distinguishes three basic sub-types of antonym, and we shall adopt his terminology here. 'Polar antonyms' are the fully-fledged antonyms which satisfy Lyons' test, where neither comparative implies its positive; pairs like GOOD/BAD, where one comparative does imply the positive, Cruse terms 'overlapping antonyms'. In a third sub-type, 'equipollent antonyms', both comparatives imply their corresponding positives; examples include English NICE/NASTY, HOT/COLD and SWEET/SOUR. Note that these antonyms fail Lyons' test in both directions (cf. Cruse, 1976: 287):

(14)a. *Sunday was nicer than Saturday, but both days were nasty.
b. *Sunday was nastier than Saturday, but both days were nice.

Cruse further points out differences in denotation between the three groups: whereas polar antonyms are typically evaluatively neutral and objectively descriptive, overlapping antonyms have an evaluative polarity (commendatory vs. deprecatory) as part of their meaning; equipollent antonyms denote subjective sensations or emotions, or evaluations based on subjective reactions.

With these distinctions in mind, let us return to the Japanese terms. We will follow Cruse's definition of antonyms in terms of the following properties (cf. 1986: 204):

(i) Antonyms are fully gradable.
(ii) A pair of antonyms denote degrees of some variable property.
(iii) When more strongly intensified, the members of a pair move in opposite directions along the relevant scale.
(iv) The terms of a pair do not strictly bisect their domain: there is a neutral region on the scale between the areas covered by the two terms.

These properties are illustrated for OISHII/MAZUI and UMAI/MAZUI in the following. Firstly, with regard to gradability, the following collocations with degree adverbials commonly occur in the spoken language and are readily accepted by the informant:

(15) sugoku *oishii/umai/mazui* 'very good-tasting/bad-tasting'

(16) totemo *oishii* 'very good-tasting'

(17) kekkoo *oishii/umai* 'pretty good-tasting'

(18) nakanaka *oishii/umai* 'quite good-tasting'

(19) maamaa *oishii/umai* 'passably good-tasting'

(20) kanari *mazui* 'pretty bad-tasting'

(21) ammari *oishiku* nai 'not very good-tasting'

(22) taishite *umaku* nai 'not very good-tasting'

(23) sonna-ni *mazuku* nai 'not that bad-tasting'

(24) chittomo *oishiku/umaku* nai 'not in the least good-tasting'

(25) zenzen *oishiku/umaku* nai 'not good-tasting at all'

The following additional collocations occur in the secondary spoken language data (*Fujin Kooron*, May 1988 (= *FK*), and Shibata and Ishige, 1983 (= SI)):

(26) saikoo-ni *oishii* (*FK*: 316) 'tremendously good-tasting'

(27) hijoo-ni *oishii* (*FK*: 320) 'very good-tasting'

(28) mechamecha *umai* (*FK*: 321) 'incredibly good-tasting'

In general, differences in adverbial collocations of OISHII and UMAI reflect stylistic factors. Grading is also exemplified in comparative sentences; Japanese adjectives lack comparative forms, and comparative constructions commonly involve postpositional elements such as YORI 'than', HODO 'as' and/or the noun HOO 'alternative'. Cf. the following examples from the secondary data:

(29) Akashi no tai yori mo Nihonkai no ga zettai *umai* tte iimasu ne. (*FK*: 313)
'They say that Japan Sea bream are definitely better-tasting than those from Akashi, don't they.'

(30) Koogan o tabetasa ni korosu to kikimashita. Sore hodo *oishii* tte. (*FK*: 316)
'I heard that they kill (goats) out of wanting to eat their testicles. They say they're that good-tasting.'

(31) ...soozoo shita yori haruka ni *umakatta.* (*FK*: 316–17)
'...it was far better-tasting than I imagined.'

(32) Arai hoo ga *oishii* n janai. (*FK*: 320)
'Isn't it that they taste better when (the sea) is rough?'

(33) ...aratte tabeta hoo ga *umai* koto o mitsuke... (SI: 45)
'...they found that they tasted better if you washed them before eating them...'

(34) Nihon no baai hie yori kome no meshi no hoo ga *umai* ni kimatte ite mo... (SI: 245)
'Though it goes without saying that, in Japan's case, rice tastes better than rice-millet...'

Cf. also the following collocations with the adverb MOTTO '(even) more' accepted by the informant:

(35) motto *oishii/umai/mazui*
'(even) more good-tasting/bad-tasting'

Thus:

(36) Are mo *mazukatta* kedo kore wa motto *mazui* yo.
THAT/integrator/BAD-TASTING, past/BUT/THIS/isolator/MORE/BAD-TASTING, non-past/emphasizer
'That was bad-tasting too, but this is even worse.'

Superlative collocations typically involve the adverb ICHIBAN 'most, to the greatest degree':

(37) Osashimi ga ichiban *oishii*. (*FK*: 316)
'Raw (goat) meat tastes best.'

(38) Nihonkai tte sakana de ittara ichiban *umai*. (*FK*: 320)
'The Japan Sea is best if you're talking about fish.'

(39) Kawahagi de tsukutta no ga ichiban *umai*. (*FK*: 321)
'The (dish) made with *kawahagi*-fish tastes best.'

The postulation of a single denotational dimension ('palatibility') for these terms is supported by their use by psychologists and food scientists in hedonic scales designed to measure subjects' judgements of the palatability of foods. Ishima (1983: 122) illustrates a seven-point scale with the following designations, where intensified terms clearly move in opposite directions:

(40) hijoo-ni *oishii* 'very good-tasting'
kanari *oishii* 'pretty good-tasting'
yaya *oishii* 'slightly good-tasting'
dochira to mo ienai 'cannot say either'
yaya *mazui* 'slightly bad-tasting'
kanari *mazui* 'pretty bad-tasting'
hijoo-ni *mazui* 'very bad-tasting'

The scale also exemplifies additional collocational possibilities with degree adverbials, at least for contexts of this type. While this scale juxtaposes OISHII and MAZUI, a more detailed scale dating from 1959 and reproduced by Ishima (1983: 119) features UMAI and MAZUI. The co-occurrence of both proposed pairs of antonyms is further illustrated by the following examples from the secondary data. First, OISHII and MAZUI:

(41) ...*oishikereba* waratte iru, *mazukereba* shimezura... (SI: 128)
'...if it tastes good (the face on the chart) is laughing, if it tastes bad it's frowning...'

(42) Sore ni, tabeta koto no nai mono wa, saisho wa *mazui* to kanjiru koto ga ooi to omoimasu. Koora o saisho ni nonda toki wa kusurikusai to kanjite mo, nonde iru uchi ni narete *oishiku* naru to iu koto mo atte... (SI: 137)
'Also, I think that things one hasn't eaten before often strike one as bad-tasting at first. Though cola might taste like medicine the first time, as you drink it you tend to get used to it and find it tastes good...'

(43) Nihonjin ni wa kome no aji no *oishii*, *mazui* ga taihen komakaku wakaru. (Koono, 1974: 105)
'The Japanese are very discriminating with regard to the palatibility (lit. good-tastingness and bad-tastingness) of the taste of rice.'

For UMAI and MAZUI:

(44) Amate no hito ga tsukutta konnyaku wa *umai*. Karate no hito no wa dokoka *mazui*. (*FK*: 318)
'*Konnyaku* paste made by people with "sweet hands" tastes good. That made by people with "salty hands" somehow tastes bad.'

(45) Hyooka suru ningen ga motte iru kitai ga manzoku saretara '*umai*' to iu shi, uragirareta toki wa '*mazui*' to iu deshoo. (SI: 136)
'Surely, when the expectations of the evaluating person are met he will say "That tastes good" and when they are not he will say "That tastes bad".'

The following designation is included in the UMAI/MAZUI hedonic scale mentioned above from Ishima (1983: 119):

(46) *umaku* mo *mazuku* mo nai
'neither good-tasting nor bad-tasting'

This illustrates the neutral region or 'no man's land' characteristic of antonymous scales. Other similar expressions, for both OISHII/MAZUI and UMAI/MAZUI, are also readily accepted by the informant:

(47) *Oishii* to wa ienai ne. *Mazuku* mo nai kedo.
GOOD-TASTING, non-past/quotation/isolator/SAY, potential, negative, non-past/tag/BAD-TASTING/integrator/negative, non-past/BUT
'You couldn't say it tastes good. It doesn't taste bad either, though.'

(48) *Mazuku* wa nai kedo taishite *umaku* mo nai.
BAD-TASTING/isolator/negative, non-past/VERY/GOOD-TASTING/integrator/negative, non-past
'It doesn't taste bad but it doesn't taste very good, either.'

Finally, we must return to the question of the sub-type of antonymy to which these terms belong. We will follow Cruse in proceeding from the semantic relation between comparative and positive uses: do explicitly comparative uses of these terms (*X wa Y yori oishii* 'X tastes better than Y', *X no hoo ga oishii* 'X tastes better', etc.) imply the corresponding positive use (*X wa oishii* 'X tastes good') or not? The following sentences were tested for acceptability with the informant:

(49)a. *Ryoohootomo *mazui* kedo dotchi ka ttara kotchi no hoo ga *oishii/umai.*
BOTH/BAD-TASTING, non-past/BUT/WHICH ONE/question/quotation, conditional/THIS ONE/genitive/ALTERNATIVE/subject/GOOD-TASTING, non-past
'Both taste bad, but if anything this one tastes better.'

b. *Ryoohootomo *oishii/umai* kedo dotchi ka ttara kotchi no hoo ga *mazui.*
'Both taste good, but if anything this one tastes worse.'

As indicated, both were rejected, (49a) after slight hesitation and in favour of (49a') or (49a"):

(49)a.' Ryoohootomo *mazui* kedo dotchi ka ttara kotchi no hoo ga mada mashi da.
'Both taste bad, but if anything this one is better.'

(49)a." Ryoohootomo *mazui* kedo dotchi ka ttara kotchi no hoo ga mada ii.
'Both taste bad, but if anything this one is better.'

with MASHI-NA ('better (than a poor alternative)') (49a') or II 'good' (49a") replacing OISHII/UMAI. Similar reactions were obtained with other Japanese native speakers.

The results indicate that OISHII/MAZUI and UMAI/MAZUI are equipollent antonyms, like English NICE/NASTY (cf. (14a, b)) and unlike GOOD/BAD. (The acceptability of (49a") also indicates that II behaves differently from OISHII and UMAI in this respect.) According to Cruse, equipollent adjectives denote subjective reactions or evaluations based thereon: OISHII, UMAI and MAZUI are clearly evaluative, and in 4.2 we have noted their grammatical behaviour as subjective adjectives.

Cruse proposes a general explanation for the differences between the subtypes of antonym in terms of the degree of autonomy of the properties concerned:

> Many of the differences between the groups can be given an intuitively satisfying explanation if we assume that for overlapping and equipollent antonym pairs (but not polars) the properties denoted by each of the members of a pair are conceived as being quasi-autonomous. That is to say, whereas SHORTNESS, for example, is no more and no less than the absence of LENGTH, GOODNESS and BADNESS, CLEVERNESS and DULLNESS, and HOTNESS and COLDNESS are to some extent independent properties. At a more abstract level, of course, there is a common underlying gradable property for both members of a pair (otherwise they could not be antonyms): MERIT for *good:bad*, TEMPERATURE for *hot:cold*, etc. But at a more superficial level, the properties have a certain independence. We can thus say that whereas a single scale underlies a pair of polar antonyms, there are two scales underlying a pair of overlapping or equipollent antonyms. (1986: 210)

It is at least intuitively plausible to regard 'good-tastingness' and 'bad-tastingness' as distinct sensation qualities, comparable to hotness and coldness. Following Cruse's diagrammatic representation of these latter dimensions for English (1986: 211), we may represent the scales involved in the case of the Japanese terms as follows:

(50)

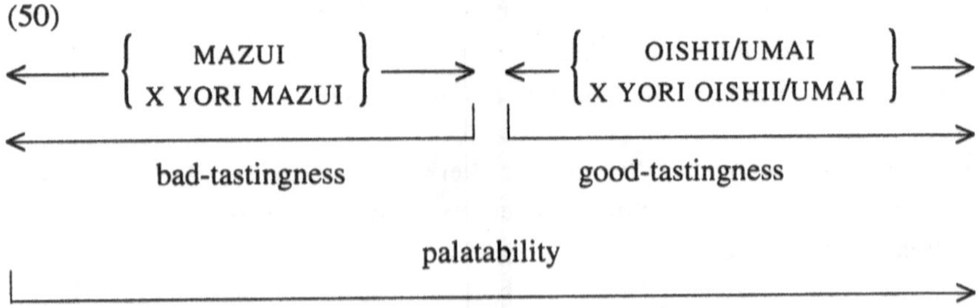

As shown by the unacceptability of (49a) and (49b), something which tastes worse than something else (X YORI MAZUI) cannot simultaneously be good-tasting (OISHII or UMAI), and vice versa, so that the scales do not overlap (cf. Cruse, 1986: 211).

4.5 Stylistic factors

In this section we examine the stylistic properties, firstly, of OISHII and UMAI, and subsequently of MAZUI and of the derivatives of these three terms.

In 4.4 we have analysed OISHII and UMAI as cross-varietal synonyms: the two items are stylistically differentiated, and in 2.4.2 we have indicated that UMAI, together with certain other basic items relating to the domain of food and eating, forms part of the 'masculine' vocabulary. As such, we expect UMAI to be used typically by male speakers in informal situations, while OISHII will be used elsewhere (i.e. by female speakers generally, and by male speakers in more formal situations).[8]

As an obvious consequence of this characterization, we will expect to find regular sex-based switching between UMAI and OISHII in informal conversations involving male and female speakers. Cf. the following constructed examples ((M) = male speaker, (F) = female speaker):

(51) (M) Kono momo wa *umai*?
(=(1)) (F) Ee, suimi ga atte *oishii*.
THIS/PEACH/isolator/GOOD-TASTING, non-past
YES/TARTNESS/subject/BE PRESENT, conjunctive/GOOD-TASTING, non-past
'Are these peaches good?'
'Yes, they have a tartness and taste good.'

(52) (F) *Oishii*?
(M) Un, *umai*.
GOOD-TASTING, non-past
YES/GOOD-TASTING, non-past
'Is it good?'
'Yes, it is.'

These examples are accepted by the informant as natural exchanges: the female speaker uses OISHII, while the male speaker opts for UMAI in view of the informality of the situation. Similarly, the following constructed example, from a dictionary entry for OISHII, is interpreted as involving the comment of a male speaker and a female response:

(53) (M) Tanigawa no mizu ga tsumetakute *umai* ne.
(F) Ee, totemo *oishii* wa. (Kyooiku Gijutsu Kenkyuujo, 1986: 104)
'Mountain stream water is cold and tastes good, doesn't it.'
'Yes, very good.'

The following are attested examples from the secondary spoken data:

(54) (F) Maguro nanka, yaite *oishii*-desu ka.
(M) *Umai.* (*FK*: 319)
'What about tuna – is it good grilled?'
'It is.'

(55) (M) ...soozoo shita yori haruka ni *umakatta*.
(F) Kitai shitenakatta kara, *oishikatta* no to chigaimasu? (*FK*: 316–17)
'...it was far better-tasting than I imagined.'
'Was it perhaps good-tasting because you weren't expecting it?'

Let us also note here the exchange in the final scene of Ozu Yasujiroo's well-known film *Ochazuke no Aji* ('The Flavour of Green Tea over Rice'), which features the husband and wife eating *ochazuke* together:

(56) (F) *Oishii* wa.
(M) *Umai* ne.
'It tastes good.'
'It does, doesn't it.'

In the written language, Koono 1974 and 1977 are works of popularizing non-fiction, written in the impersonal (*de aru*) style. OISHII is used throughout, with the exception of the following three examples of UMAI:

(57) Omake-ni, Ishida Mitsunari wa Tokugawakata ni tsukamari shokei no tame hikitaterarete iku tochuu, *umai* mizu ga nomitai to itta. (Koono, 1974: 94)
'On top of this, Ishida Mitsunari was captured by the Tokugawa forces and, while being led away to execution, said that he wished to drink some good-tasting water.'

(58) Aroma wa, koohii o tatete iru toki ni tachinobotte kuru kaori no yoo ni, shokuhin kara tachinobotte kita kaori ni yotte, *umasoo* da na to kanjisaseru kaori de aru. (Koono, 1974: 162)
'Aroma is the fragrant smell which arises from foods and makes one think "This smells good", like the smell from brewing coffee.'

(59) ...toku ni tsuyoi shigeki wa nai ga, miryoku no aru no ga ochazuke no aji to suru nara, koku ga atta, *umakatta* to inshoo ni nokoru no ga biifusuteeki no aji de aru. (Koono, 1977: 18)
'...if one were to characterize the taste of *ochazuke* (= green tea poured over boiled rice) as attractive, though without any particularly strong stimulus, then the taste of beef steak is one which one remembers as having been rich-tasting, as having been good-tasting.'

In all three cases UMAI (in the case of (58), the derived evidential UMASOO-NA; cf. 4.2) occurs within quotative clauses, marked by the quotative particle TO. (57) reports the male speech of Ishida Mitsunari, while (58) and (59) record private impressions or soliloquizings, presented as male by the (male) author.

While the linguistic characterization of UMAI as being restricted to masculine speech is clear, this does not mean that, in real life, the item is never used by females. Such uses may occur when directly quoting male speech, for example. Less trivially, women may also use UMAI in cases where, for whatever reason, they wish to project a masculine image. As it happens, attested examples of this are found in the *Fujin Kooron* round-table discussion data, where the female participant uses UMAI several times at the beginning of the discussion. These also occur in quotative clauses, and in some cases are interpretable as referring to words or thoughts of (non-specific) male speakers; the following example, however, reports the speaker's own reactions:

(60) Kore wa katsuji ni shite hoshiku nai kedo, karugamo, kawaii na to omou to dooji ni, *umasoo* da naa tte... (*FK*: 313)
'I wouldn't want this to appear in print but (when I see a) *karugamo*-duck, at the same time as thinking "How cute!", (I also think) "How good-tasting it looks"...'

This example is marked as eccentric for the informant, attesting to the existence of the stylistic norm. Although the speaker (an internationally renowned violinist) subsequently reverts to the use of OISHII throughout the remainder of the discussion, she clearly abandons here the persona of a Japanese female.

In addition to its use by male speakers, we have also referred to the informal nature of UMAI: typically, 'masculine' items occur in male informal speech. Morphological evidence for the informality of UMAI vis-a-vis OISHII is provided by the absence of an extra-formal form (cf. 2.4.2) for the former:

(61)

Informal	*umai*	*oishii*
Formal	*umai-desu*	*oishii-desu*
Extra-formal	**umoo-gozaimasu*	*oishuu-gozaimasu*

(All forms are non-past, predicative.)[9]

At the same time, UMAI does occur, as illustrated, in the formal form (*umai-desu*). Once again, attested examples occur in the *Fujin Kooron* data:

(62) Moto wa nan da tte kikasarete, gembutsu miserareru to gyotto shimasu kedo, *umai*-desu ne, ano suupu. (*FK*: 315)
'It gives you a fright when you're told what it's made from and shown the actual article, but it's good, you know, that soup.'

(63) Okinawa de ichiban suki-na no wa mozuku. *Umai*-desu, are wa. (*FK*: 316)
'What I like best in Okinawa is *mozuku* (seaweed). It's good, that.'

In this respect, UMAI is less restricted than other masculine items in this domain: the verbs KUU 'eat' with human subjects and (HARA GA) HERU 'feel hungry' (cf. 2.4.2), for example, do not readily occur in formal forms (*kuimasu*, (*hara ga*) *herimasu*). On the scale of masculinity, UMAI is thus less excessively 'macho' than these neighbouring items (cf. Chapter 2, note 12).[10]

We may conclude that UMAI is typically restricted to use by male speakers, in at least relatively informal circumstances. Unlike women, who (normally) use only OISHII, men thus have two lexical options and we would predict that they will use OISHII in situations of relative formality and/or where they do not wish to project masculinity. The dynamic nature of the factors involved is clear from the *Fujin Kooron* data, where each of the male speakers uses both UMAI and OISHII in the same discussion – the overall tone of which may be characterized as good-humoured and relatively informal. The following figures show the frequency of use of the two items (including derivatives) by the participants:

(64)

	UMAI	OISHII
Senior male	25	10
Junior male	2	2
(Female	3	20)

(Total length of the discussion is approximately 5,000 running words, with relative contribution rates to this total from the three participants in the order of 4 (senior male): 1 (junior male): 2 (female).)

The figures clearly suggest a correlation between seniority and frequency of use of UMAI vis-a-vis OISHII, but they also indicate the presence of additional relevant pragmatic factors. One of these may be empathy with the female addressee (cf. Ide, 1982: 380; Mizutani and Mizutani, 1987: 77): several uses of OISHII by the senior male occur in comments or questions directed to the female participant. Another may concern the nature of the topic and may explain the apparent preference for OISHII in collocation with KUUKI 'air' noted in 4.3: to express appreciation of the quality of air is to display a certain refinement of sensitivity which sits incongruously with the simultaneous expression of masculinity through robust speech.

Collocation is a further source of evidence of stylistic status, and we will in general expect stylistic congruity to be maintained in syntactic constructions, and indeed over longer stretches of discourse. What constitutes a 'stretch' in the relevant sense here is clearly a complex matter, involving the workings of a variety of pragmatic factors; for present purposes let us simply say that we will predict the use of UMAI, rather than OISHII, in the company of the masculine items MESHI 'boiled rice, meal', KUU 'eat' and HARA GA HERU 'feel hungry'.[11] Thus the following examples from the *Fujin Kooron* data illustrate UMAI co-occurring with KUU and MESHI:

(65) ...asoko ni *umai* sushiya ga atte, soko de kutta toki, nante *umai* sakana daroo to omoimashita. (*FK*: 313)
'...there's a good *sushi* place there, and when I had (snapper) there I thought what a good-tasting fish it was.'

(66) Nezumiiro no mizu, demo kuttara *umai*, mechamecha *umai*. (*FK*: 321)
'It was (covered with) dark grey water, but when you ate it it tasted good, incredibly good.'

(67) Saisho, konna namagusai mono kueru ka to omou n da kedo, *umai*. Saigo ni, shiroi meshi ni sore o kakete... (*FK*: 321)
'At first you wonder if you can eat something so fishy-smelling, but it tastes good. Finally you pour it over white rice and...'

Similarly the following example, from a dictionary entry for UMAI (Kyooiku Gijutsu Kenkyuujo, 1986: 88), features UMAI co-occurring with HARA GA HERU:

(68) Aa, hara ga hetta, okaasan, nanka *umai* mono wa nai?
'Gee, I'm hungry, mum, isn't there anything good (to eat)?'

Other examples in the *Fujin Kooron* data (all spoken by male speakers) all support the prediction that KUU and MESHI call up UMAI. The converse, however, does not hold, as shown in the following example where UMAI co-occurs with TABERU rather than with KUU:

(69) ...hitori sanjuubi gurai tabemashita yo, *umakute*. (*FK*: 320)
'...we each ate about thirty (sardines), they were (so) good-tasting.'

This bears out our earlier conclusion, based on morphological evidence, that UMAI is less strongly marked for masculinity than these other terms. The relative disposition of these items with respect to the stylistic parameter of masculinity may thus be represented as follows:

(70)

<table>
<tr><td rowspan="3">Masculinity (+ ↑ … ↓ −)</td><td rowspan="2">UMAI</td><td>KUU</td></tr>
<tr><td rowspan="2">TABERU</td></tr>
<tr><td>OISHII</td></tr>
</table>

Our discussion has assumed that MESHI/GOHAN stylistically parallel KUU/TABERU. While the *Fujin Kooron* material does not provide a comparable range of examples for these terms, informant reactions certainly indicate that this is the case: as stated in 4.3, UMAI collocates acceptably with both GOHAN and MESHI, OISHII with GOHAN alone (cf. *Kono gohan umai/oishii ne*, *Kono meshi umai/ *oishii ne* 'This (boiled) rice tastes good, doesn't it.').

As a diachronic footnote to the discussion of the stylistic contrast of UMAI and OISHII, we note here that OISHII is historically much the later term, having its origins in *nyoobookotoba*, the euphemistic vocabulary of court ladies-in-waiting which first arose around the beginning of the fifteenth century. These items were particularly plentiful in the domains of food and clothing, and were frequently adopted into the language of females more widely. According to Hachiya, UMAI 'good-tasting' was originally replaced in *nyoobookotoba* by the adjective ISHII, which in turn developed into OISHII (through prefixation of O-; cf. 2.4.2); OISHII appears to have been restricted to female speech until late in the last century, since when its range has widened as we have indicated here (1983: 32–7). The Linguistic Atlas of Japan shows many peripheral dialects still possessing only a single item, corresponding to UMAI; whereas UMAI is present throughout the country, OISHII is distributed over central Honshu and adjoining areas, bespeaking historically later diffusion from a centre in this region (Kokuritsu Kokugo Kenkyuujo, 1974: 86–8).

Turning to MAZUI 'bad-tasting', we have seen in 4.4 that this term co-occurs as an antonym with both OISHII and UMAI. This suggests that MAZUI is stylistically unrestricted. This view is supported by Nishio (1972: 103), who states that the stylistic opposition between OISHII and UMAI is neutralized in this item, and appears to be universally held.

Nevertheless, evidence of the following kind suggests that MAZUI is subject to some restrictions on usage:

(71) (F) Doo?
(M) *Mazui* naa.
(F) *Oishiku* nai tte.
'How is it (i.e. the food)?'
'It tastes bad.'
'He says it doesn't taste good.'

(71) is interpreted as a natural exchange in an informal situation where a female speaker asks a male his opinion of the food he is eating, and then reports his unfavourable response to a third person. The following is a similar attested example:

(72) (M) Sakki made *oishikatta* kedo ima ni natte *mazuku* natta n da yo.
(F) *Oishiku* naku natchatta?
'(This wine) tasted good until just now but has now become bad-tasting.'
'Has it become not good-tasting?'

In both these examples it is noticeable that MAZUI used by male speakers is 'echoed' by negative forms of OISHII by female speakers. Other cases of the use of *oishiku nai*, etc. where we might expect forms of MAZUI are found, for example, in the secondary data from Shibata and Ishige, which reports the proceedings of a conference in the personal formal (*desu/-masu*) style:

(73) ...amami ga tsuyoku naru to *oishiku* naku naru to iu iken mo dete kimasu. (SI:117)
'...we also find some opinions that, as the sweetness becomes stronger, it becomes not good-tasting.'

(74) ...mata doobutsushiboo wa doomyakukooka no moto da to iwareru to, buta no abura ga doku mitai ni miete *oishiku* naku natte shimau to iu wake desu. (SI:117)
'...and, when we are told that animal fats are the cause of hardening of the arteries, pig fat is seen as harmful and becomes not good-tasting.'

(75) Kodomo ga yasai o *oishiku* nai to itte tabenai no wa... (SI: 118)
'The reason why children say vegetables don't taste good and don't eat them...'

(The author of these examples is male.) At the same time, some contributors to these proceedings use MAZUI, as seen in earlier examples; furthermore, Koono (1974, 1977), noted above as avoiding UMAI except in a small number of clearly male contexts, freely uses MAZUI throughout, the only exception noted being the following example which contains both MAZUI and *oishiku nai*:

(76) Mata...shokuhinzairyoo no *mazuku* natta koto ni yotte, iya de mo kaori de gomakasanai to, ryoori ga *oishiku* naku natta. (Koono, 1974: 165–6)
'Moreover...due to the fact that foodstuffs have become bad-tasting, the cuisine has become not good-tasting unless one resorts to camouflage by means of flavourings.'

The examples appear to indicate an avoidance of MAZUI particularly by female speakers and occasionally, although not to the same extent as with UMAI, elsewhere in more formal contexts.[12] Given that this is so, the question arises: what happens in antonymous contexts in such cases? What do female speakers say, for example, when they wish to describe an item of food as tasting neither good nor bad? The following response was elicited from the informant:

(77) (M) *Oishii*, sono sakana?
(F) Maa, toku-ni *oishiku* mo nai kedo *oishiku* naku mo nai.
GOOD-TASTING, non-past/THAT/FISH
WELL/ESPECIALLY/GOOD-TASTING/integrator/negative, non-past/BUT/GOOD-TASTING/negative/integrator/negative, non-past
'Does it taste good, that fish?'
'Well, it doesn't taste especially good but it doesn't taste not good, either.'

The response in (77) is accepted by other native speakers, and is marked stylistically as being spoken by a female. It is a highly unusual sentence, containing a negative and a double negative form of a single lexical item (OISHII) in a construction typically filled by negative forms of a pair of antonyms (OISHII/MAZUI: *oishiku mo nai kedo mazuku mo nai* 'doesn't taste good but doesn't taste bad, either'). Analogous sentences for other pairs of antonyms are not tolerated. Thus, in (78):

(78) *Kono kuruma wa toku-ni ookiku mo nai kedo ookiku naku mo nai.
'This car isn't especially big but it's not not big, either.'

we must have *chiisaku mo nai* 'is not small, either', with a form of the antonym CHIISAI 'small' in the second clause. The same is true, indeed, of UMAI 'good-tasting':

(79) *Maa, toku-ni *umaku* mo nai kedo *umaku* naku mo nai.
'Well, it doesn't taste good but it doesn't taste not good, either.'

where we must use *mazuku mo nai* 'is not bad-tasting, either', with a form of MAZUI. The explanation for the occurrence of the unusual response in (77) appears to be that the only candidate for the status of antonym to OISHII, namely MAZUI, is stylistically excluded in this context and the normal linguistic pattern overruled: women favour negative forms of OISHII over MAZUI even to the extent where, as in (77), semantically the negative must be interpreted as a contrary (admitting of a neutral region between the two extremes of the scale) rather than as a complementary (where neutral regions are excluded).[13] This constitutes a solution to a situation which is probably unique in the language: while other examples of stylistic differentiation in antonyms are found, the normal state of affairs is for both terms to have variants at the different stylistic levels, so that no 'gap' occurs.

This avoidance of MAZUI appears to be particularly strong where the taste of prepared dishes is involved (and, by implication, the competence of a preparer). Thus the informant reports that she feels no compunction about using MAZUI with reference to medicines, for example; cf. the following attested example from a magazine, where a mother is writing about her child:

(80) Nigakute *mazui* kusuri desu ga...nonde kuremasu.
'It's bitter, bad-tasting medicine but (the child) takes it.'

In 5.4.2.5 below, MAZUI is indeed elicited from the informant with reference to substances such as sea-water and *shibugaki* persimmons (cf. Chapter 5, note 19). We must recognize that there is thus a difference in range, and ultimately in the nature of the motivation for the restriction, from the case of UMAI. The effect, however, is the same, i.e. to exclude MAZUI from female speech, at least where reference is to orthodox items of diet; if we amend our earlier diagram ((70)) to incorporate the usage of MAZUI for these contexts, the result is as follows:

(81)

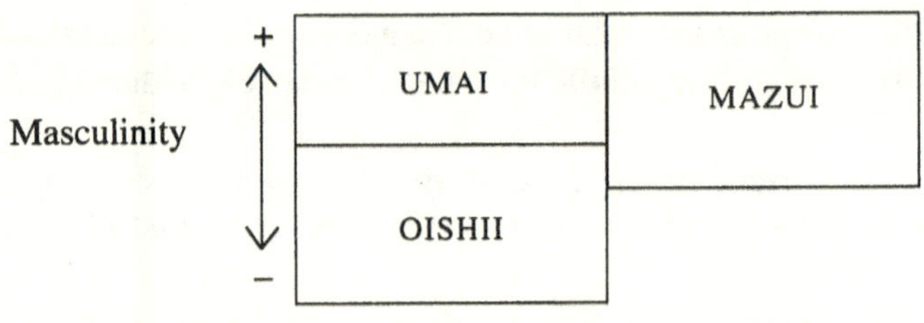

In terms of usage, MAZUI overlaps with UMAI and, partly, with OISHII. At the 'soft' end of the scale where female speakers typically operate, however, taste evaluation is expressed in terms of the single item OISHII and its negation.

Finally we consider the stylistic properties of derivatives of evaluative taste terms, focussing on OISHII and UMAI. The following derivatives have been mentioned (cf. 4.2):

(82)

	Adjective (-SOO-NA)	Verb	Noun (-SA)	Noun (-MI)
OISHII	OISHISOO-NA	OISHIGARU	OISHISA	
UMAI	UMASOO-NA	UMAGARU	UMASA	UMAMI

In general, these derived items share the stylistic characteristics of their base lexemes. The exception is the derived noun UMAMI; the nominalizing suffix -MI is much less productive than -SA and, as is clear from the above table, there is no formal opposition in this 'slot' since the corresponding derivative is absent in the case of OISHII.

That UMAMI is not stylistically restricted in the same way as UMAI is clear from the following elicited example:

(83) Kore wa samma no hontoo no *umami* ga gozaimasen ne.
THIS/isolator/SAMMA (fish)/genitive/REALITY/genitive/
GOOD-TASTINGNESS/subject/BE PRESENT, extra-formal,
negative, non-past/tag
'This hasn't got the real tastiness of a *samma*-fish, has it.'

where UMAMI collocates with the extra-formal verb form *gozaimasen* (cf.2.4.2). In everyday language UMAMI commonly occurs in the expression UMAMI GA ARU 'good-tastingness is present, tasty' or (as in (83)) its stylistic variants, chiefly applied to fish and seafood and denoting the natural tastiness associated with these items; the expression contrasts with the term OOAJI-NA '(large and) insipid', applied to tasteless specimens which have overgrown their gastronomic prime, and its evaluative nature is clear. There is also a technical use of UMAMI in the language of food scientists, where it is likewise stylistically unrestricted. It functions here as a descriptive term, applied to the taste of various acids:

(84) ...shokuhingakuteki ni iu to, *umami* to iu no wa aminosankei no aji, kakusan no aji o iimasu ga... (Ishima, 1983: 115)
'...in terms of food science, *umami* refers to the taste of amino acids and nucleic acids...'

Its technical nature in this use is clearly evident from the experiments of O'Mahony and Ishii, who report that the term was only used by personnel involved in the manufacture and tasting of monosodium glutamate to refer to the taste of that substance (1986: 165, 170). As well as MSG (derived from *kombu* 'kelp'), O'Mahony and Ishii mention substances derived from bonito flakes (*katsuobushi*) and black mushrooms (*shiitake*) as types of *umami* seasonings (1986: 169; cf. also Nagao, 1982: 34).

4.6 Extralingual meaning

In the previous sections we have examined the intralingual semantic relations of OISHII, UMAI and MAZUI, and we have also considered their stylistic properties. In this section we address the question of how these terms relate more directly to the world outside language. Our discussion will focus on two main aspects: the extent of their extralingual range, and the question of their subjectivity. This latter is a difficult and controversial area in the analysis of evaluative terms: do such terms denote objective, public properties of stimuli, or express subjective, private reactions of experiencers – or both?

Before considering the set as a whole, let us make the preliminary point concerning OISHII and UMAI that, given their intralingual analysis as synonyms, we will naturally expect these terms to have identical extralingual meanings (although they may differ in other aspects of use). Morita (1989: 197) appears to challenge this when he notes that, whereas OISHII is applied to pleasant tongue tastes, the meaning of UMAI extends to depth of flavour. This view contrasts with that of Nishio (1972: 103), who maintains that in the modern standard language OISHII and UMAI contrast only in terms of style, and with that of Tokugawa and Miyajima (1972: 54), who claim that these terms are semantically 'completely identical' (*mattaku onaji*). On closer inspection Morita's only supporting example employs the derived noun UMAMI 'tastiness', with which, as we have seen in 4.5, there is no opposition with a comparable derivative from OISHII, and he presents no evidence involving the basic adjectives themselves. We will therefore continue to hold the view that OISHII and UMAI are identical in extralingual meaning; any differences in applicability, we will assume, are to be related to their stylistic differences (cf. the case of KUUKI 'air' considered in 4.5). It may well be, for example, that UMAI, rather than OISHII, is commonly applied to the taste of whisky. However, this may be readily explained in terms of the fact that whisky is typically drunk by men, and thus tends to be commended in masculine terms; in more formal contexts, or where the speaker is female, OISHII is perfectly appropriate.[14]

With regard to the extralingual range of the set, we have in 4.4 used the label 'palatability' for the underlying property invoked by the three terms. An important question concerns the extent of the range of this property: as we have seen (1.1), the taste experience is a complex one, involving the interaction of various senses. The following attested example clearly suggests that the present terms range over all parameters of this experience, and thus involve 'taste' in the widest sense.

(85) Kono pirafu wa ammari *oishiku* nai. Aji wa ii kedo betabeta-shiteru.
THIS/PILAFF/isolator/(NOT) VERY/GOOD-TASTING/negative, non-past/TASTE/isolator/GOOD, non-past/BUT/STICKY, non-past
'This pilaff isn't very good. The taste is okay, but it's glutinous.'

The example indicates that an item can have a satisfactory 'taste' (AJI) in some narrower sense but still fail to be OISHII if other parameters (here, dryness, consistency) are inadequate. This wide range is further illustrated by the variety of terms which may occur in slot Y in the sentence-frame *(Kono X wa) Y oishii/ umai/ mazui* '(This X) is Y and (thereby) good-tasting/bad-tasting'. Sentences of this type are commonly used in response to the question *Sono X wa doo?* 'How is that X?', used in general enquiries regarding palatability. Cf. the following examples accepted by the informant:

(86)	Sono momo wa doo? THAT/PEACH/isolator/HOW 'How is that peach?'	Suimi ga atte *umai*. TARTNESS/subject/BE PRESENT, conjunctive/ GOOD-TASTING, non-past 'It has a tartness and tastes good.'
(87)	Sono nori wa doo? 'How is that laver?'	Kaori ga yokute *oishii*. AROMA/subject/GOOD, conjunctive/GOOD-TASTING, non-past 'It's aromatic and tastes good.'
(88)	Sono tsukemono wa doo? 'How are those pickles?'	Piritto-shitete *oishii*. PIQUANT, conjunctive/GOOD-TASTING, non-past 'They're piquant and taste good.'

(89) Sono kamaboko wa doo? — Bosoboso-shitete *mazui*.
DRY, conjunctive/BAD-TASTING, non-past
'How is that fish-sausage?' — 'It's dry and tastes bad.'

(90) Sono sembei wa doo? — Paripari-shitete *umai*.
CRUNCHY, conjunctive/GOOD-TASTING, non-past
'How are those rice crackers?' — 'They're crunchy and taste good.'

(91) Sono biiru wa doo? — Namanurukute *oishiku* nai.
LUKEWARM, conjunctive/GOOD-TASTING/negative, non-past
'How is that beer?' — 'It's warm and doesn't taste good.'

(86)–(91) contain forms of the expressions SUIMI GA ARU 'has a tartness', KAORI GA II 'aromatic', PIRITTO-SHITE IRU 'sharp', BOSOBOSO-SHITE IRU 'dry', PARIPARI-SHITE IRU 'crunchy' and NAMANURUI 'lukewarm' which, by virtue of their occurrence in this frame, may be characterized as taste terms in the widest sense; in this case, the terms range over the parameters of gustation, olfaction, pain, texture, hardness and temperature, respectively. Let us also note here that examples such as the following highlight extralingual differences of intensity between MAZUI 'bad-tasting' on the one hand and *oishiku nai/umaku nai* 'not good-tasting' on the other:

(92) Sono ocha wa doo? — Kaori ga nakute *oishiku* nai/*umaku* nai/**mazui*.
AROMA/subject/BE PRESENT, negative, conjunctive/...
'How is that (green) tea?' — 'It has no aroma and doesn't taste good/*tastes bad.'

In (92) we see that the predication of certain taste characteristics, here the lack of aroma expected from green tea, may be compatible with the denial of OISHII/UMAI 'good-tasting' but incompatible with the assertion of MAZUI 'bad-tasting'. MAZUI thus denotes a stronger degree of unpalatability than that involved in simple lack of aroma; this may indeed extend to inedibility, as illustrated in the common collocation *mazukute taberarenai* (BAD-TASTING, conjunctive/EAT, potential, negative, non-past) 'is bad-tasting and is (thereby) inedible'. Syntagmatic relations between descriptive taste terms and OISHII/UMAI/MAZUI are further examined within a more restricted range in 5.4.2.5.

As indicated above, the question of the subjectivity of evaluative terms is controversial. Lehrer (1983: 41–4) summarizes various philosophical approaches to items of this type, and suggests that terms such as GOOD, as used in the evaluation of wine, involve a combination of expressive and descriptive functions: while they are basically used to commend, this commendation is based on more or less generally accepted objective criteria. We will take a somewhat similar view here with respect to the Japanese terms, but with stronger emphasis on their primary subjective function of expressing personal affective reactions of the speaker; at the same time, we must be mindful of the fact that individual taste reactions and preferences are subject to the channelling influence of culturally accepted criteria (as expressed, for example, in taste norms: cf. 1.4), and these may be of pragmatic importance in the interpretation of discourse dealing with taste.

Let us return to the linguistic characteristics of the Japanese terms, and in particular to their membership, noted in 4.2, of the sub-class of subjective adjectives. Extralingually, adjectives of this sub-class express emotional or perceptual states and reactions, as indicated by the following representative members:

> HOSHII 'find desirable, want', URAYAMASHII 'find enviable', ZANNEN-NA 'find regrettable', KUYASHII 'find vexing', KANASHII 'feel/find sad', URESHII 'feel pleased/find pleasing', ARIGATAI 'feel grateful/find welcome', MENDOOKUSAI 'find troublesome', URUSAI 'find importunate', KOWAI 'feel frightened/find frightening', OSOROSHII 'find horrifying', KAWAII 'find lovable', SABISHII 'feel lonely', NATSUKASHII 'find nostalgic', HAZUKASHII 'feel ashamed, shy', MEIWAKU-NA 'find incommoding', OKASHII 'find amusing, strange', OMOSHIROI 'find interesting, diverting', FUSHIGI-NA 'find odd', NEMUTAI 'feel sleepy', MABUSHII 'find dazzling', ITAI 'feel/find painful', KAYUI 'feel itchy', KUSUGUTTAI 'find ticklish', OMOTAI 'find heavy', SAMUI 'feel cold', ATSUI 'feel hot'

Subjective adjectives are associated with an experiencer and, typically, a stimulus, commonly expressed by noun phrases basically marked by the subject particle GA; glosses in the list above are cast in verbal form to reflect this. In discourse, both the experiencer and the stimulus may be unexpressed; they may also be expressed by noun phrases marked by focus particles such as WA and MO, which replace GA.

In examining the Japanese terms, we will focus on three types of evidence for their status as subjective adjectives. According to Nishio, the most characteristic feature of adjectives of this type is that they are directly predicable of first-person experiencers (1972: 25–8). While these remain unexpressed in most situations, the following example, readily accepted by the informant, shows that OISHII is naturally predicated of such an argument when the context demands:

(93) (A) Kono pirafu wa ammari *oishiku* nai ne.
(B) Soo? Watashi wa kekkoo *oishii* kedo ne.
THIS/PILAFF/isolator/(NOT) VERY/GOOD-TASTING/negative, non-past/tag/SO/I/isolator/QUITE/GOOD-TASTING, non-past/BUT/TAG
'This pilaff isn't very good, is it?'
'Oh? I find it quite good.'

This may be contrasted with the case of an objective adjective such as the colour term AKAI 'red':

(94) (A) Kono ringo wa zuibun akai ne.
(B) *Soo? Watashi wa ammari akaku nai kedo.
(Soo? Watashi wa ammari akai to omowanai kedo)
'These apples are really red, aren't they?'
'Oh? I don't find them that red.'

As indicated in (94), AKAI is not directly predicable of the experiencer and a verb such as OMOU 'think, consider' must be employed. Nishio himself seems somewhat reluctant to class the present terms as subjective, but the example he suggests of a possible exchange among diners (1972: 105) is readily accepted by the informant:

(95) (A) *Oishii.*
(B) Boku mo *oishii.*
'(It')s good.'
'I think so too.'

Nishio also notes that all three terms form derived verbals in -GARU; as mentioned in 4.2, this serves as a useful morphological indicator of the sub-class.

The process converts subjective adjectives into verbs denoting the external manifestation of the internal states or reactions involved: 'It [-GARU] changes verbals of internal feeling into those of outward manifestation of internal feeling' (Kuno, 1973: 84). -GARU verbs may be intransitive or transitive (or both) (cf. note 5), and may be glossed accordingly as 'show signs of feeling Adj' or 'show signs of finding (something) Adj'. Nishio gives the following constructed example for MAZUGARU (1972: 105):

(96) *Mazugatte* sukoshi mo tabenai.
'(He) shows signs of finding (the food) bad-tasting and won't eat any of it.'

The following similar example with OISHIGARU is readily accepted by the informant:

(97) (A) Appurupai wa?
(B) Kodomotachi ga *oishigatte* zembu tabechatta.
'What about the apple pie?'
'The children loved it (lit. showed signs of finding it good-tasting) and ate it all.'

Finally, evidence for the subjective status of these terms comes from constructions of the following kind involving adverbial forms:

(98) Taihen *oishiku* itadakimashita.
EXTREMELY/GOOD-TASTING, adverbial/CONSUME (object-honorific), formal, past
'I ate/drank (your food, etc.) with relish (i.e. in a finding-good-tasting manner).'

(98) is a formula used to express appreciation of a previous gift of food and may be compared with the following similar examples:

(99) Taihen omoshiroku haidoku-shimashita.
EXTREMELY/INTERESTING, adverbial/READ (object-honorific), formal, past
'I read (your book, etc.) with great interest (i.e. in a finding-interesting manner).'

(100) Taihen natsukashiku haiken-shimashita.
EXTREMELY/NOSTALGIC, adverbial/SEE (object-honorific), formal, past
'It brought back great memories to see (your photograph, etc.) (i.e. I saw it in a finding-nostalgic manner).'

(98)–(100) exemplify a construction involving a verb denoting an act of perception, an (unexpressed) stimulus object, and the adverbial form of an adjective expressing the internal reaction produced. As indicated in the glosses, the adjectives clearly require a subjective interpretation. Comparable examples with objective adjectives are rejected:

(101) *Taihen akaku haiken-shimashita.
'I saw (your picture, etc) very redly (i.e. in a finding-red manner).'

(102) *Taihen utsukushiku haichoo-shimashita.
'I heard (your composition, etc) in a finding-beautiful manner.'

AKAI 'red' and UTSUKUSHII 'beautiful' are objective adjectives. Although this construction is particularly common in polite formulae of the type illustrated, more general examples with OISHII and TABERU 'eat'/NOMU 'drink' are found in the secondary data:

(103) Hachimitsu o mazeru to *oishiku* nomeru no wa... (Koono, 1977: 112)
'The reason why one can drink (lemon juice) enjoyably (i.e. in a finding-good-tasting manner) when you mix honey (with it)...'

(104) Soko de...katai niku o nantoka shite *oishiku* taberu koto no kufuu ga hitsuyoo ni natta no de aru. (Koono, 1974: 53)
'At this point...it became necessary to devise some way of eating tough meat enjoyably (i.e. in a finding-good-tasting manner).'

(105) ...ika-ni tanoshiku, mata, *oishiku* taberu ka to iu koto ni me ga mukanai koto de aru. (Koono, 1974: 135)
'...it means that one pays no attention to the question of how one eats with enjoyment and relish (i.e. in a finding-enjoyable and finding-good-tasting manner).'

(106) ...sakana o *oishiku* taberu koto ga dekiru. (Koono, 1974: 157)
'...one can eat fish enjoyably (i.e. in a finding-good-tasting manner).'

With second- or third-person experiencers, the plain adjective must be replaced by an expression referring to outward manifestation: derived evidential adjectives (in their adverbial form: *oishisoo-ni*, etc.) are commonly used here. Cf. the following natural collocations:

(107) *oishisoo-ni* taberu 'eat with apparent relish (i.e. in an apparently finding-good-tasting manner)'
umasoo-ni taberu 'eat with apparent relish'
mazusoo-ni taberu 'eat with apparent distaste'
omoshirosoo-ni yomu 'read with apparent interest'
natsukashisoo-ni miru 'look at with apparent nostalgia'

This construction is wide-ranging and, once again, the need for a subjective interpretation of the adjectives concerned is evident. Once again, objective adjectives are excluded:

(108) *akasoo ni miru 'look at in an apparently finding-red manner'
*utsukushisoo ni kiku 'look at in an apparently finding-beautiful manner'

In summary, then, subjective adjectives in Japanese, which express psychological states and reactions of various kinds, are basically predicated of first-person experiencers; with second- and third-person experiencers, reference must be made instead to the external manifestations of the experience involved. In this section we have presented ample evidence that the evaluative taste terms participate in the grammar of subjective adjectives. In 4.4 we have seen that the terms behave as equipollent antonyms, and that these are noted by Cruse as referring to 'distinctly subjective sensations or emotions (e.g. HOT:COLD, HAPPY:SAD), or evaluations based on subjective reactions, rather than on "objective" standards (e.g. NICE:NASTY, PLEASANT:UNPLEASANT)' (1986: 208).[15] Just as English NICE differs from GOOD in this respect, so OISHII and UMAI differ from II 'good' (cf. 4.4). II and OISHII also relate differently to the adjective SUKI-NA 'like, have a liking for':

(109) Kore wa ii wain da to omou kedo, watashi wa ammari suki janai.
THIS/isolator/GOOD, non-past/WINE/copula, non-past/quotation/THINK, non-past/BUT/I/isolator/(NOT) VERY/LIKE/ copula, negative, non-past
'I think this is a good wine, but I don't like it very much.'

(110) *Kore wa *oishii* wain da to omou kedo, watashi wa ammari suki janai.
'I think this is a good-tasting wine, but I don't like it very much.'

(109) shows that one may accept that a wine is good in a general sense (II) and simultaneously disclaim personal liking; cf. Lehrer (1983: 42), who makes a similar point for GOOD in English. With OISHII, however, this is not possible, as illustrated in (110); OISHII, like SUKI-NA, reflects personal taste.

Based on the discussion in this section, we propose the following characterization of the extralingual meaning of OISHII, UMAI and MAZUI:

(111) 'I find the taste (in the widest sense) of X pleasant (OISHII/UMAI)/ unpleasant (MAZUI).'

Semantically, then, these adjectives are subjective in the sense that their basic extralingual function is to express a pleasurable or unpleasurable reaction to the taste experience.[16] At the same time, we must keep in mind the fact, noted in 1.4, that individual judgements of taste are influenced by cultural values, with respect both to what items are regarded as food and to why. What materials or ingredients are considered good to eat (or drink)? In what manner are they prepared? And on what basis is their quality appraised? The role of culture in providing answers to these questions is readily apparent. Clearly, specific expressions of appreciation such as *Kono ringo wa oishii* 'This apple tastes good' are to be interpreted in relation to the relevant reference-set (cf. Lehrer, 1983: 42), and thus to a greater or lesser extent against background cultural norms: in general, a given specimen of food is evaluated in terms of what it is supposed to taste like, and such matters are largely culture-dependent. Despite the essentially subjective meaning of OISHII, UMAI and MAZUI, therefore, we must recognize that they will generally be applied against an implicit background of Japanese norms, and that a familiarity with these norms will thus be important for a full interpretation of evaluative discussions of food in Japanese. At a general level we have referred to the restricted gustatory range of the traditional Japanese diet (cf. 1.4): thus Koono refers to its narrow range of seasonings (soy sauce, soybean paste (*miso*), or salt alone), to its unelaborate cooking methods, and to its characteristic smooth, rather than coarse, textures (1977: 101, 159). He also refers to the high degree to which culinary norms are shared throughout the culture (although now under threat from the encroachment of processed foods) (1977: 174–7), and this doubtless contributes to the ready verbalization of norms for specific items of diet noted in 1.4.

In short, subjectivism here needs to be tempered by a recognition of the influence of cultural values, so that a fuller characterization of the extralingual meaning of these terms will often be the following:

(112) '(X meets/fails to meet relevant Japanese taste norms and) I find the taste of X pleasant/unpleasant.'

However the bracketed component constitutes a pragmatic implicature rather than a semantic implication. Individual preferences remain, and examples such as the following are perfectly acceptable:

(113) Kotchi no ringo wa Nihon no ringo yori suppai kedo, *oishii* ne.
'The apples over here are sourer than Japanese apples, but they're good, aren't they.'

4.7 Extended meanings

In this section we briefly address the question of extended meanings in relation to Japanese evaluative taste terms. As outlined in 2.3, our present approach leads us to discuss extensions in terms of the modifications which are effected upon the semantic relations, both intralingual and extralingual, established for the basic system. Of particular interest are cases where two or more basic terms participate in a given extension, with paradigmatic relations held more or less constant; such cases reflect the linguistic reality of the basic system, and may provide further evidence as to the status of terms within it.

With regard to semantic range the most striking feature of the present three terms is that, whereas OISHII 'good-tasting' is virtually restricted to the domain of taste, the pair UMAI 'good-tasting' and MAZUI 'bad-tasting' are used, roughly as opposites, in two other domains in the modern language (cf. Morita, 1989: 194–7). In the first, which we may label 'skilfulness', UMAI is equivalent to 'good (at), well executed', MAZUI to 'poorly executed'; [17] as indicated, UMAI has the wider range here, being applied to performers as well as to products (cf. *umai/mazui e* 'skilful/unskilful painting', *umai/*mazui ekaki* 'skilful/unskilful painter'). Interestingly, UMAI is stylistically marked, as colloquial, in this domain vis-a-vis JOOZU-NA 'skilful' (cf. Tokugawa and Miyajima, 1972: 54); the term HETA-NA 'unskilful' serves to fill the semantic gap left by MAZUI in application to performers (*heta-na ekaki* 'unskilful painter'). Secondly, they occur in a domain where UMAI may be glossed as 'favourable, felicitous' and

MAZUI as 'unfavourable, infelicitous'. While clearly contrastive in meaning, they are in fact rarely parallel in distribution here: UMAI, for example, has a common adverbial use in this domain (cf. *umaku iku* '(things) go well/as one would wish'), but this is not found with MAZUI. A characteristic collocation of UMAI in this meaning is with the noun HANASHI 'talk, proposal, affair'; *umai hanashi* refers to a proposal or idea that promises to bring benefit to those concerned, and this collocation serves as the clear basis for a unique, and apparently quite recent, extension of OISHII outside the domain of taste, illustrated in the following attested example:

(114) Pitchaa ni wa *oishii* hanashi desu ne.
'It (i.e. the proposal to remove the 'lucky zone' from certain baseball grounds, thus rendering home runs more difficult to achieve) is a favourable idea for pitchers, isn't it.'

As previously noted, none of these terms is directly extended to the domain of smell: a good food-smell is *umasoo-na/oishisoo-na nioi* rather than **umai/ *oishii nioi* (cf. 4.3). In this respect the terms differ from English items such as DELICIOUS, as well as from certain descriptive taste terms in Japanese (cf. 6.4).

Outside taste, therefore, we essentially find the pair UMAI/MAZUI applied in the two domains described: while the tightness of the intralingual opposition is somewhat relaxed, both domains retain the polarity found in taste, with UMAI as the positive term, and the pair clearly occupies an important place in the overall evaluative vocabulary of Japanese.[18] On structural grounds, their wide occurrence might be taken to reflect the status of UMAI/MAZUI, rather than OISHII/MAZUI, as the central opposition in taste. However, it seems clear that the restriction of OISHII in this respect is due to its relatively recent historical development (cf. 4.5) rather than to any lack of centrality as a taste term, where we have seen that, at least with respect to style, UMAI rather than OISHII is the marked term in the modern language.

The precise historical details of the extension of UMAI and MAZUI are not clear: as far as taste is concerned, UMAI is certainly long established in this domain (Katoo, 1966; Hachiya, 1983: 32), while MAZUI is a relative latecomer, dating from the eighteenth century (Hachiya, 1983: 36). Synchronically, however, there is no doubt that taste is the primary focus of the opposition in the modern language: it is here that the distribution of the terms is most clearly parallel and the relation of antonymy most sharply delineated.

5 *Descriptive taste terms in Japanese (I)*

In the following two chapters we treat a set of descriptive taste terms in Japanese, defined by our delimitation procedure as constituting the lexical field of AJI 'taste'. We will see that, extralingually, the members of this field range over a domain wider than that of 'taste' in the scientific sense (i.e. gustation), but narrower than that of 'taste' when it is equated with the whole of the ingestion experience, including parameters such as texture, consistency and temperature.

The present chapter is organized as follows. In 5.1 the lexical field is delimited, and the grammatical characteristics of the member terms are described in 5.2. In 5.3 the initial field is subdivided into the lexical systems AJI I, AJI II, and AJI III, and the semantic structure of AJI I, the largest system, is analysed in 5.4. The semantic structure of the remaining systems, and other semantic aspects of the field, are described in Chapter 6.

5.1 Delimitation

As outlined in 3.1, the general strategy adopted in this study toward the investigation of lexical fields is, by asking appropriate questions, to determine what lexical fields and systems, with what members, are present in the language, and subsequently to examine the semantic relations which hold among members. That is to say, the question of what a lexical item means is approached by considering what basic question(s) it occurs in answers to, what other items also occur in such answers, and how the items are semantically related to each other and to aspects of extralinguistic reality.

Our initial task is thus to determine how the Japanese ask questions about taste; in particular, for our present purposes, we wish to discover a natural way of seeking descriptive information on taste qualities in Japanese. As a point of general methodology, one may seek to elicit such questions in various ways, in this case, for example, by uttering a sentence referring to a culturally unfamiliar item of diet such as the following:

(1) Kinoo papaiya to iu mono o tabeta yo.
YESTERDAY/PAPAYA/quotation/SAY, non-past/THING/object/EAT, past/emphasizer
'Yesterday I ate something called a papaya.'

A natural response to such an utterance would be:

(2) Papaiya tte donna mono?
PAPAYA/quotation/WHAT SORT OF/THING
'What kind of a thing is a papaya?'

Appropriate responses to this question, in turn, would provide information concerning the size, colour, shape, taste, consistency, etc. of the papaya, so that (2) is clearly too wide in scope to provide an ideal question-frame for our particular purposes. By obstinately refusing to give information concerning the taste of the item in response to (2), however, the investigator may provoke the question:

(3) Donna aji ga suru?
WHAT KIND OF/TASTE/subject/BE PRESENT, non-past
'What kind of a taste does it have?'

(3) is clearly narrower in scope than (2) in that appropriate responses to it form a proper subset of appropriate responses to (2): information concerning size, colour, shape, etc. is rejected in this case. We therefore use (3) as the basis for our first question-frame (A), aimed at delimiting descriptive taste terms (i.e. AJI terms) in Japanese:

(A) X wa donna aji ga suru?[1]
X/isolator/WHAT KIND OF/TASTE/subject/BE PRESENT, non-past
'What kind of a taste does X have?'

Slot X in (A) is typically filled by generic nouns denoting food substances, i.e. nouns denoting items of diet in the strict sense, as well as substances such as medicines, sea-water, etc. which are ingested more or less commonly.

Our aim at this point is thus to isolate the set of lexical items available in the Japanese language system for the description of taste qualities on a general level, and the question-frame must be used to elicit as wide a range of responses as possible, to allow us to approximate as closely as possible to the complete membership of the field. The lexical items elicited in response to frame (A) are seen in the following question-and-answer pairs:

(A1)	Meron wa donna aji ga suru?[2]	*Amai.*
	MELON/isolator/WHAT KIND OF/ TASTE/subject/BE PRESENT, non-past	SWEET[3], non-past
	'What kind of a taste do melons have?'	'They're sweet.'
(A2)	Nihonshu wa donna aji ga suru?	Chotto *amai.*
	SAKE/...	SOMEWHAT/SWEET, non-past
	'What kind of a taste does *sake* have?'	'It's somewhat sweet.'
(A3)	Momo wa donna aji ga suru?	*Amazuppai.*
	PEACH/...	SWEET-SOUR, non-past
	'What kind of a taste do peaches have?'	'They're sweet-sour.'
(A4)	Mikan wa donna aji ga suru?	*Amazuppai.*
	MANDARIN/...	SWEET-SOUR, non-past
	'What kind of a taste do mandarins have?'	'They're sweet-sour.'
(A5)	Remon wa donna aji ga suru?	*Suppai.*
	LEMON/...	SOUR, non-past
	'What kind of a taste do lemons have?'	'They're sour.'
(A6)	Koogyoku wa donna aji ga suru?	Chotto *suppai.*
	JONATHAN (APPLE)/...	SOMEWHAT/SOUR, non-past
	'What kind of a taste do Jonathans have?'	'They're somewhat sour.'
(A7)	Sukiyaki wa donna aji ga suru?	*Amakarai.*
	SUKIYAKI (meat and vegetables cooked in soy sauce and sugar)/...	SWEET-PUNGENT, non-past
	'What kind of a taste does *sukiyaki* have?'	'It's sweet-pungent.'
(A8)	Tamanegi wa donna aji ga suru?	*Karai.*
	ONION/...	PUNGENT, non-past
	'What kind of a taste do onions have?'	'They're pungent.'

(A9)	Shiomizu wa donna aji ga suru? SALT-AND-WATER SOLUTION/... 'What kind of a taste does a salt-and-water solution have?'	*Karai.* PUNGENT, non-past 'It's pungent.'
(A10)	Karashina wa donna aji ga suru? MUSTARD-PLANT/... 'What kind of a taste does mustard-plant have?'	*Piritto karai.* PIQUANTLY/PUNGENT, non-past 'It's piquantly pungent.'
(A11)	Umi no mizu wa donna aji ga suru? SEA/genitive/WATER/... 'What kind of a taste does sea-water have?'	*Shiokarai.* SALT, non-past. 'It's salt.'
(A12)	Shiomizu wa donna aji ga suru? SALT-AND-WATER SOLUTION/... 'What kind of a taste does a salt-and-water solution have?'	*Shoppai.* SALTY, non-past 'It's salty.'
(A13)	Kobucha wa donna aji ga suru? KOBUCHA (tea made from a type of seaweed)/... 'What kind of a taste does *kobucha* have?'	Chotto *shoppai.* SOMEWHAT/SALTY, non-past 'It's somewhat salty.'
(A14)	Konagusuri wa donna aji ga suru? MEDICAL POWDER/... 'What kind of a taste do medical powders have?'	*Nigai.* BITTER, non-past 'They're bitter.'
(A15)	Biiru wa donna aji ga suru? BEER/... 'What kind of a taste does beer have?'	*Horonigai.* PLEASANTLY BITTER, non-past 'It's pleasantly bitter.'
(A16)	Shibugaki wa donna aji ga suru? SHIBUGAKI (a type of persimmon)/... 'What kind of a taste do *shibugaki* have?'	*Shibui.* ASTRINGENT, non-past 'They're astringent.'

(A17)	Momo wa donna aji ga suru?	*Amakute suimi ga aru.*
	PEACH/...	SWEET, conjunctive/ TARTNESS/subject/BE PRESENT, non-past
	'What kind of a taste do peaches have?'	'They're sweet and have a tartness.'
(A18)	Sakkarin wa donna aji ga suru?	*Amai* kedo *nigami ga aru.*
	SACCHARINE/...	SWEET, non-past/BUT/ BITTERNESS/subject/ BE PRESENT, non-past
	'What kind of a taste does saccharine have?'	'It's sweet but has a bitterness.'
(A19)	Umi no mizu wa donna aji ga suru?	*Karakute* chotto *nigami ga aru.*
	SEA/genitive/WATER/...	PUNGENT, conjunctive/ SOMEWHAT/ BITTERNESS/subject/ BE PRESENT, non-past
	'What kind of a taste does sea-water have?'	'It's pungent and has a slight bitterness.'
(A20)	Desugita ocha wa donna aji ga suru?	*Nigakute shibumi ga aru.*
	OVERINFUSE, past/GREEN TEA/...	BITTER, conjunctive/ ASTRINGENCY/subject/ BE PRESENT, non-past
	'What kind of a taste does overinfused green-tea have?'	'It's bitter and has an astringency.'
(A21)	Goboo wa donna aji ga suru?	*Akuppoi.*
	BURDOCK/...	HARSH, non-past
	'What kind of a taste does burdock have?'	'It's harsh.'
(A22)	Goboo wa donna aji ga suru?	*Aku ga aru.*
	BURDOCK/...	HARSHNESS/subject/ BE PRESENT, non-past
	'What kind of a taste does burdock have?'	'It has a harshness.'

(A23)	Koohii wa donna aji ga suru? COFFEE/... 'What kind of a taste does coffee have?'	*Kaori ga ii.* AROMA/subject/GOOD, non-past 'It's aromatic.'
(A24)	Hoojicha wa donna aji ga suru? ROASTED TEA/... 'What kind of a taste does roasted tea have?'	*Koobashii.* FRAGRANT, non-past 'It's fragrant.'
(A25)	Chiizu wa donna aji ga suru? CHEESE/... 'What kind of a taste does cheese have?'	*Kusami ga aru.* SMELLINESS/subject/ BE PRESENT, non-past 'It's strong-flavoured.'
(A26)	Koi wa donna aji ga suru? CARP/... 'What kind of a taste does carp have?'	*Dorokusai.* MUDDY-SMELLING/ non-past 'It's muddy-flavoured.'
(A27)	Tomato wa donna aji ga suru? TOMATO/... 'What kind of a taste do tomatoes have?'	*Sammi ga atte* chotto *aokusai.* SOURNESS/subject/ BE PRESENT, conjunctive/ SOMEWHAT/GRASSY-SMELLING, non-past 'They have a sourness and are somewhat grassy-flavoured.'
(A28)	Banana wa donna aji ga suru? BANANA/... 'What kind of a taste do bananas have?'	*Amakute kaori ga ii.* SWEET, conjunctive/ AROMA/subject/GOOD, non-past 'They're sweet and aromatic.'

(A29)	Ninjin wa donna aji ga suru?	Maa, ninjin no aji da kedo. Chotto *amami ga aru.*
	CARROT/...	WELL/CARROT/genitive/ TASTE/copula, non-past/ BUT/SOMEWHAT/ SWEETNESS/subject/BE PRESENT, non-past
	'What kind of a taste do carrots have?'	'Well, they taste like carrots – they have a slight sweetness.'

(A30)	Tako wa donna aji ga suru?	Betsu-ni aji ga nai yo.
	OCTOPUS/...	(NOT) PARTICULARLY/ TASTE/subject/BE PRESENT, negative, non-past/emphasizer
	'What kind of a taste does octopus have?'	'It doesn't have any taste in particular.'

(A31)	Shake wa donna aji ga suru?	Shake no aji to shika ienai-daroo ne.
	SALMON/...	SALMON/genitive/ TASTE/quotation/ EXCEPT/SAY, potential, negative, presumptive/tag
	'What kind of a taste does salmon have?'	'I suppose you could only say it tastes like salmon.'

Of these responses, (A1)–(A28), and the second half of the response in (A29), can be summarized as consisting either of the form of a single taste expression or of a sequence of the forms of two such expressions, with the additional presence of a modifying element (CHOTTO 'somewhat') in certain cases. Although question-frame (A) was used with a wide variety of nouns filling slot X, no terms were elicited other than those represented in the responses to (A1)–(A29), although, needless to say, other combinations of these terms were found to occur.

The above sample of data includes cases (e.g. (A3) and (A17)) where the same noun appears with different terms occurring in responses on different occasions: such cases may be of important heuristic value in exploring the semantic structure of the field. (A30) and (A31) are also included as constituting appropriate responses, though of a different semantic order. The response to (A30) illustrates what might be termed the 'zero' case in taste, analogous to the use of the term COLOURLESS in relation to the English colour terms. (A31), on the other hand, illustrates a characteristic of the domain of taste which appears to have no parallel in colour: in the latter domain, given the set of English colour terms plus the term COLOURLESS, it is in principle possible to answer any appropriate question of the form *What colour is X?* using only those terms, singly or in combination, with or without the additional use of modifiers such as DARK, LIGHT, etc.; i.e. in this sense, colour terms constitute an exhaustive lexical continuum. (A31) indicates that this is not the case with taste terms, in that it demonstrates that there are substances whose taste cannot be subsumed within the range of taste terms plus the 'zero' expression (cf. Kuipers, 1984: 91). This has certain consequences for the statement of intralingual relations, notably the fact that, whereas the denial of a colour term implies the assertion of the disjunction of the remaining set of colour terms at the same hierarchical level (*X is not red* implies *X is green/black/yellow/... /colourless*, etc.), the denial of a taste term will have no such closed set of implied disjunctions.

5.2 Grammatical properties

In this section we examine the main morphological and syntactic characteristics of the items elicited in (A1)–(A29). We also comment briefly on the status of neighbouring items which were not elicited by frame (A).

Grammatically the elicited items can be divided into four groups, as follows:

(i)

AMAI 'sweet'
SUPPAI 'sour'
AMAZUPPAI 'sweet-sour'
KARAI 'pungent'
AMAKARAI 'sweet-pungent'
SHIOKARAI 'salt'
SHOPPAI 'salty'
NIGAI 'bitter'
HORONIGAI 'pleasantly bitter'
SHIBUI 'astringent'
AKUPPOI 'harsh'

(ii)

PIRITTO 'piquantly'

(iii)

AMAMI GA ARU 'has a sweetness'
SUIMI GA ARU 'has a tartness'
SAMMI GA ARU 'has a sourness'
NIGAMI GA ARU 'has a bitterness'
SHIBUMI GA ARU 'has an astringency'
AKU GA ARU 'has a harshness'
KUSAMI GA ARU 'strong-flavoured'

KOOBASHII 'fragrant'	(iv)
DOROKUSAI 'muddy-flavoured'	KAORI GA II 'aromatic'
AOKUSAI 'grassy-flavoured'	

Group (i) consists of *i*-adjectives (cf. 4.2), while the single term in group (ii) is an adverb. Groups (iii) and (iv) consist of phrasal expressions of the form noun + subject particle + verb (ARU) or *i*-adjective (II), respectively.

Morphologically, of the fourteen adjectives in group (i), four have stems which are incontrovertibly simple. Of the remainder, seven have clearly composite stems, and three are of doubtful status. The structure of the stems in this group may be illustrated as follows:

ama-	('sweet')
kara-	('pungent')
niga-	('bitter')
shibu-	('astringent') (cf. *shibu* 'astringent juice of persimmons, etc.')
shoppa-	('salty'): (?) *sho-* (<*shio* 'salt') + *-ppa*
suppa-	('sour'): (?) *su* 'vinegar'/*su-* 'sour' + *-ppa*
koobashi-	('fragrant'): (?) *koo* 'fragrance, incense' + *-bashi*
amazuppa-	('sweet-sour'): *ama-* 'sweet' + *suppa-* 'sour'
amakara-	('sweet-pungent'): *ama-* 'sweet' + *kara-* 'pungent'
shiokara-	('salt'): *shio* 'salt' + *kara-* 'pungent'
horoniga-	('pleasantly bitter'): *horo-* 'to a pleasant degree' + *niga-* 'bitter'
akuppo-	('harsh'): *aku* 'lye' + *-ppo* '-ish'
dorokusa-	('muddy-flavoured'): *doro* 'mud' + *kusa-* 'smelly'
aokusa-	('grassy-flavoured'): *ao-* 'blue, green' + *kusa-* 'smelly'

The illustrated analyses for SHOPPAI and SUPPAI, in particular, must be regarded as uncertain, and Nishio (1983: 109), for example, regards both stems as simple. In the first place, the proposed suffix *-ppa* is of very low productivity, being restricted to these adjectives and elsewhere to a few nouns such as HAPPA 'leaf', HARAPPA 'open field, open land'; all these items are colloquial in style (cf. Backhouse, 1983). The reduction of *shio* to *sho* is also restricted to SHOPPAI, so that the relation is synchronically tenuous, although historically accurate: *shoppai* results from the composite *shio-hayui*, itself a development from an earlier *shiwa-hayui* 'salty' (lit. 'lip-tingling') (cf. Tokugawa, 1979: 217–18; Hachiya, 1983: 30). Diachronically, *suppai* is thought to have been formed by analogy with *shoppai*, replacing earlier SUI and (dialectal) SUKKAI 'sour' (Kokuritsu Kokugo Kenkyuujo, 1966b: 26). As for KOOBASHII, the proposed analysis is supported by the orthography, but the suffix *-bashi* is virtually unique to this item; diachronically, the form results from the composite *ka-guwashii* 'fragrant' (lit. 'aroma-beautiful'). Amongst the clearly composite items, *horo-* (in HORONIGAI) occurs as an initial element in items such as HOROYOI 'slight intoxication, tipsiness', and *-ppo* (in AKUPPOI) is a productive adjectivalizing suffix, typically associated with pejorative meaning and colloquial style (cf. MIZUPPOI 'watery, insipid', KODOMOPPOI 'childish', YASUPPOI 'cheap-looking', etc.).

PIRITTO in group (ii) is formed from the mimetic base *piri-* by the addition of the adverbial particle *to* with accompanying gemination of /t/.[4] Mimetic bases enter into a variety of phonological and morphological processes, and PIRITTO relates formally to items such as PIRIRI(-TO) 'piquantly' (a less colloquial form), PIRIPIRI 'smartingly', HIRIHIRI 'burningly (of skin, tongue)', etc.

The nouns in group (iii) phrases, with the exception of AKU 'harshness', are derivatives formed with the nominalizing suffix -MI (cf. 4.2). The prevalence of -MI nominalizations, and their typical occurrence in phrases of the present kind with the verb ARU, appear to be particularly characteristic of the domain of taste and thus provide some formal support for the linguistic reality of the present field.[5] Three of the derivatives, AMAMI 'sweetness', NIGAMI 'bitterness' and SHIBUMI 'astringency', are derived from the corresponding adjectives in group (i), while KUSAMI 'strong flavour' is based on the adjective KUSAI 'smelly' which occurs as the second element in the group (i) terms DOROKUSAI 'muddy-flavoured' and AOKUSAI 'grassy-flavoured'. SUIMI[6] 'tartness' is apparently derived from the adjective SUI 'sour', which is now largely archaic in the standard language vis-a-vis SUPPAI but survives in many dialects in Western Japan (cf. Kokuritsu Kokugo Kenkyuujo, 1966b: 25–6), while SAMMI 'sourness' is derived from the noun SAN 'acid'. Note that the expected derivative from SUPPAI, namely SUPPAMI, is not used by the informant (though listed in many dictionaries). The remaining noun in group (iii), AKU 'harshness', is morphologically simple and appears as the base of AKUPPOI 'harsh' in group (i).

In group (iv), the noun KAORI 'pleasant smell, aroma' is also morphologically simple.

In addition to the -MI nominalizations indicated, all the adjectives in group (i) produce abstract nouns in -SA (cf. 4.2). Several also produce derived verbs in -GARU, a morphological indicator of subjective adjectives (cf. 4.2, 4.6): Martin's list of adjectives which undergo -GARU verbalisation (1975: 361–5) includes SUPPAI 'sour' (acceptability rating: 75–100 per cent), for which he gives the following example:

(4) Kodomo ga natsumikan o *suppagaru*.
'Children find the citron too sour for them.' (1975: 363)

He also lists NIGAI 'bitter' and SHIBUI 'astringent' (both with acceptability ratings of 50–75 per cent) but, surprisingly, omits KARAI 'pungent' despite the fact that sentences such as the following containing KARAGARU are quite natural:

(5) Kyoo karee o tsukutta kedo minna *karagatte* tabete kurenakatta.[7]
'I made some curry today, but everyone found it too hot and refused to eat it.'

The subjective status of descriptive taste adjectives will be discussed further in 5.4.

Finally in this section we mention cases of putative taste terms which were not elicited from the informant in responses to question-frame (A). Nishio lists the adjectives EGUI (with the variant EGOI) and EGARAPPOI (variant IGARAPPOI) alongside KARAI and SHIBUI, and describes them as denoting the irritating sensation produced in the throat by the harshness (*aku*) of certain foods (1972: 98, 102); IGARAPPOI is also mentioned by Yamada, a Tokyo speaker, as being in his active taste vocabulary, although he is unable to identify precisely the taste quality it describes (1972: 31–2). Like Yamada, the informant uses IGARAPPOI, but with reference to the irritation to the throat caused, for example, by the smoke from a burning cigarette; for the comparable sensation produced by food substances, the informant employs AKUPPOI 'harsh' and AKU GA ARU 'has a harshness'.[8] Also listed by Nishio are USUAMAI ('slightly sweet') and AMATTARUI ('unpleasantly sweet') (1972: 98). USUAMAI is not used by the informant, and the only attested example cited by Nishio involves an extended use outside the domain of taste (1972: 102). AMATTARUI, by contrast, is both actively used by the informant and well attested in Nishio's data (1972: 273–4). As Nishio makes clear in his analysis, the term combines the implication of sweetness with the expression of negative evaluation; we surmise that this latter component inhibits its appearance in response to frame (A), which seeks general descriptive information on the taste of food categories.

5.3 Semantic subdivision

At this point we must consider whether the set of terms delimited as constituting the lexical field of AJI can be semantically subdivided.

For this purpose, we introduce a new question-frame (B), which can be represented simply as:

(B) X wa Y?
X/isolator/Y
'Is X Y?'

where X is filled, as in question-frame (A), by lexical items denoting general food categories, and Y by a member of the set of taste terms which have been found to occur in response to (A). For example:

(B1) Banana wa *amazuppai*?
BANANA/isolator/SWEET-SOUR, non-past
'Are bananas sweet-sour?'

yields the response:

Amazuppaku nai. *Amai.*
SWEET-SOUR/negative, non-past/SWEET, non-past
'No, they're not. They're sweet.'

In addition to examples of this kind, where the negative form of one item is followed by the positive form of another, we find cases like (B2):

(B2)	Chiizu wa *karai*?	*Karaku* nai.
	CHEESE/isolator/PUNGENT, non-past	PUNGENT/negative, non-past
	'Is cheese pungent?'	'No, it's not.'

where there is simply the negative form of one item in the response, without a following positive. With the same noun in question-frame (A), however, our data contains the response shown in the following:

(A25)	Chiizu wa donna aji ga suru?	*Kusami ga aru.*
	'What kind of a taste does cheese have?'	'It's strong-flavoured.'

Given that this is so, there is *a priori* no reason not to expect a reply to (B2) parallel to (B1), viz.:

(B2')	Chiizu wa *karai*?	**Karaku* nai. *Kusami ga aru.*
	'Is cheese pungent?'	'No, it's not. It's strong-flavoured.'

(B2'), however, is not forthcoming, and when suggested to the informant is rejected as being incoherent: the term KUSAMI GA ARU 'strong-flavoured' is felt as being out of place in this context.

We interpret this as indicating that, although KARAI 'pungent' and KUSAMI GA ARU 'strong-flavoured' both belong to the lexical field of AJI as delimited on the basis of question-frame (A), they belong to different lexical systems within it; AMAI 'sweet' and AMAZUPPAI 'sweet-sour', on the other hand, belong together in the same system, as indicated by the response to (B1). The situation may be illustrated more simplistically by the following English examples:

(6)	What do elephants look like?	They're big and grey.
(7)	Are elephants small?	No, they're not. They're big. *No, they're not. They're grey.

where BIG and GREY may be said to belong to a single lexical field at one level (the field of 'external appearance') but are distinguished at a lower level within it as belonging to the systems of 'size' and 'colour', respectively. In answer to a question about size, an item from the system of colour is out of place.

By applying this strategy over a large number of permutations of taste terms, it can be shown that they fall into three systems. The following examples will serve as typical illustrations of the procedure:

(B3)	Banana wa *kusami ga aru*?	Uun, *kusami ga nai. Kaori ga ii.*
	BANANA/isolator/SMELLINESS/ subject/BE PRESENT, non-past	NO/SMELLINESS/ subject/BE PRESENT, negative, non-past/ AROMA/subject/GOOD, non-past

	'Are bananas strong-flavoured?'	'No, they're not. They're aromatic.'
(B4)	Momo wa *amai*? PEACH/isolator/SWEET, non-past	*Amai* kedo *suimi mo aru.* SWEET, non-past/BUT/ TARTNESS/ integrator/BE PRESENT, non-past
	'Are peaches sweet?'	'Yes, they are, but they also have a tartness.'
(B5)	Koogyoku wa *suppai*? JONATHAN/isolator/SOUR, non-past	*Suppaku* wa nai kedo *sammi ga aru.* SOUR/isolator/negative, non-past/BUT/ SOURNESS/subject/BE PRESENT, non-past
	'Are Jonathans sour?'	'They're not *sour*, but they do have a sourness.'
(B6)	Momo wa *suppai*? PEACH/isolator/SOUR, non-past 'Are peaches sour?'	*Suppaku* nai. *Amazuppai.* SOUR/negative, non-past/ SWEET-SOUR, non-past 'No, they're not. They're sweet-sour.'
(B7)	Goboo wa *kaori ga ii*? BURDOCK/isolator/AROMA/ subject/GOOD, non-past	Uun, betsu-ni *kaori ga yoku* nai. NO/(NOT) PARTICULARLY/ AROMA/subject/GOOD/ negative, non-past
	'Is burdock aromatic?'	'No, not particularly.'
(B7')	Goboo wa *kaori ga ii*?	*Uun, *kaori ga yoku* nai. *Akuppoi.* (cf. (A21))
	'Is burdock aromatic?'	'No, it isn't. It's harsh.'

(B8)	Ninjin wa *akuppoi*? CARROT/isolator/HARSH, non-past 'Are carrots harsh?'	Betsu-ni *akuppoku* nai. (NOT) PARTICULARLY/ HARSH/negative, non-past 'No, not particularly.'
(B8')	Ninjin wa *akuppoi*? 'Are carrots harsh?'	**Akuppoku* nai. Chotto *amami ga aru*. (cf. (A29)). 'No, they're not. They have a slight sweetness.'
(B9)	Sakkarin wa *nigai*? SACCHARINE/isolator/ BITTER, non-past 'Is saccharine bitter?'	Soo-ne, chotto *nigami ga aru* ne. WELL/SOMEWHAT/ BITTERNESS/ subject/BE PRESENT, non-past/tag 'Well yes, it has a slight bitterness.'

As this small sample indicates, the form of the responses shows considerable variety, and some of the distinctions involved will be drawn upon below in establishing intralingual relations. The procedure was followed over the complete range of taste terms delimited, and it was found that the terms could be subdivided into three lexical systems, which we shall refer to as AJI I, AJI II and AJI III: in all cases, a question of type (B) can be answered only in terms of items within the same system. The groupings are as follows:

AJI I

AMAI 'sweet'
SUPPAI 'sour'
AMAZUPPAI 'sweet-sour'
KARAI 'pungent'
AMAKARAI 'sweet-pungent'
SHIOKARAI 'salt'
SHOPPAI 'salty'
NIGAI 'bitter'
HORONIGAI 'pleasantly bitter'
SHIBUI 'astringent'
PIRITTO 'piquantly'

AJI II

AKUPPOI 'harsh'
AKU GA ARU 'has a harshness'

AJI III

KOOBASHII 'fragrant'
DOROKUSAI 'muddy-flavoured'
AOKUSAI 'grassy-flavoured'
KUSAMI GA ARU 'strong-flavoured'
KAORI GA II 'aromatic'

AMAMI GA ARU 'has a sweetness'
SUIMI GA ARU 'has a tartness'
SAMMI GA ARU 'has a sourness'
NIGAMI GA ARU 'has a bitterness'
SHIBUMI GA ARU 'has an astringency'

5.4 The lexical system AJI I

We begin the semantic analysis with AJI I, the largest of the three systems in terms of membership. We survey extralingual meaning first (5.4.1), then proceed to examine intralingual relations (5.4.2).

5.4.1 Extralingual meaning

In this sub-section we address the broad features of the extralingual meaning of AJI I terms. The findings will be further developed in 5.4.2, in conjunction with the analysis of intralingual relations.

We begin by considering the nature of the taste qualities denoted by AJI I terms. As foreshadowed in 1.4 and 2.2, we approach this question by way of examples of substances which are typically associated with the terms concerned. In the following chart, AJI I terms are listed together with lexical items denoting food and other substances in which they are held to be typically exemplified by the informant; the entries for food substances are arranged in two columns, according to whether the associated propositions are more reasonably seen as denotational axioms or as taste norms.

(8)

AJI I term	Focal exemplars	Taste norms
AMAI 'sweet'	SATOO 'sugar' MIRIN '*mirin*' (a sweet kind of *sake* used in cooking)	BANANA 'banana'
AMAZUPPAI 'sweet-sour'		MIKAN 'mandarin' ICHIGO 'strawberry' MOMO 'peach' SUMOMO 'plum' SUBUTA 'sweet-and-sour pork'
SUPPAI 'sour'	SU '(rice) vinegar' REMON 'lemon'	UMEBOSHI 'pickled plum'
KARAI 'pungent'	KARASHI 'mustard' KAREEKO 'curry powder' KOSHOO 'pepper' WASABI '*wasabi*' (a kind of horseradish) SHOOGA 'ginger' SHOOYU 'soy sauce' SHIO 'salt'	
AMAKARAI 'sweet-pungent'		SUKIYAKI '*sukiyaki*' (a meat and vegetable dish cooked with soy sauce and sugar)
SHIOKARAI 'salt'	UMI NO MIZU 'sea-water'	
SHOPPAI 'salty'	SHIO 'salt'	KOBUCHA '*kobucha*' (tea made from a type of seaweed)

(8) *cont.*

NIGAI 'bitter'	KONAGUSURI 'medicinal powders'	BIIRU 'beer'
HORONIGAI 'pleasantly bitter'		BIIRU 'beer'
SHIBUI 'astringent'	SHIBUGAKI '*shibugaki*' (an astringent kind of persimmon)	
PIRITTO 'piquantly'	KARASHINA 'mustard-plant'	
AMAMI GA ARU 'has a sweetness'		NINJIN 'carrot'
SUIMI GA ARU 'has a tartness'		MOMO 'peach'
SAMMI GA ARU 'has a sourness'	MOKA 'mocha'	KOOGYOKU 'Jonathan'
NIGAMI GA ARU 'has a bitterness'	SAKKARIN 'saccharine'	
SHIBUMI GA ARU 'has an astringency'	BUDOO NO KAWA 'grape skins'	

Where they form part of the normal diet, substances listed as focal exemplars are in general typically used to season or garnish food, rather than being consumed on their own (cf. the substances listed against AMAI 'sweet', SUPPAI 'sour', KARAI 'pungent', and SHOPPAI 'salty'): such substances are employed in the cuisine as standard sources of the taste quality they represent, and it is particularly difficult to envisage deviant instances of them. By contrast, food items such as fruit, vegetables and cooked dishes are clearly more susceptible to variation, due to such factors as degree of ripening, length of storage, skilfulness of preparation, etc., and are thus grouped under the heading of taste norms.

Some of the individual entries require comment. It is noticeable that the three terms SHIOKARAI 'salt', NIGAI 'bitter' and SHIBUI 'astringent' have as focal exemplars substances that are not items of diet in the strict sense: sea-water, medicines, and *shibugaki* persimmons (which are considered inedible unless dried, when they become sweet). In general there are few, if any, items of normal diet which are expected regularly to possess these taste qualities, although deviant specimens may clearly do so (cf. *Kono gureepufuruutsu wa nigai* 'This grapefruit is bitter', where the general taste norm is *Gureepufuruutsu wa suppai* 'Grapefruit are sour'). With respect to the terms NIGAI 'bitter', HORONIGAI 'pleasantly bitter' and their shared exemplar beer, the norm *Biiru wa horonigai* 'Beer is pleasantly bitter' holds, broadly speaking, for people who like beer, *Biiru wa nigai* for others; this aspect of the relationship between NIGAI 'bitter' and HORONIGAI 'pleasantly bitter' will be discussed further below. Two other substances also appear against more than one taste term: salt (*shio*) (against KARAI 'pungent' and SHOPPAI 'salty') and peaches (*momo*) (against AMAZUPPAI 'sweet-sour' and SUIMI GA ARU 'have a tartness'); these cases will also be examined in the course of the analysis of intralingual relations. Finally, we may note that, according to the informant, the application of AMAKARAI 'sweet-pungent' is restricted to foods that have been cooked with sugar and soy sauce, substances which are focal exemplars for AMAI 'sweet' and KARAI 'pungent', respectively.

In general, items elicited from the informant accord well with examples of typical substances mentioned in linguistic and lexicographical treatments. These are illustrated in the following table based on (a) items employed in the Linguistic Atlas of Japan survey (Kokuritsu Kokugo Kenkyuujo, 1966a: 107–8) with the aim of eliciting taste terms in particular semantic areas, (b) items mentioned in his description of Japanese taste adjectives by Nishio (1972: 98–102), who explicitly supports illustration by typical exemplars as the most practical approach to the extralingual meaning of taste terms, and (c) items mentioned in the definitions of taste terms in three representative monolingual dictionaries: Kemboo et al. (1982) (K), Hayashi et al. (1987) (H), and Kindaichi et al. (1989) (Ki); in the case of (H), these include items mentioned in the panel entry for the taste terms AMAI, KARAI, NIGAI, SUPPAI/SUI and SHIBUI (1987: 15).

(9)

AJI I term	KKK survey	Nishio	Dictionaries
AMAI 'sweet'	SATOO 'sugar'	SATOO	SATOO (all) MITSU 'honey' (H, Ki) AME 'sweets' (H)
AMAZUPPAI 'sweet-sour'			KUDAMONO NO OOKU 'most fruits' (H)
SUPPAI 'sour'	UMEBOSHI 'pickled plum' (or AOI MIKAN 'unripe mandarins')	SU '(rice) vinegar' UMEBOSHI	UMEBOSHI (all) NATSUMIKAN '*natsumikan*' (type of orange) (K, Ki) SU (H) REMON 'lemon' (H)
KARAI 'pungent'	TOOGARASHI 'red peppers' SHIO 'salt'	TOOGARASHI KARASHI 'mustard' KAREEKO 'curry powder' DAIKON-OROSHI 'grated *daikon* radish'	TOOGARASHI (all) WASABI '*wasabi*' (all) KAREEKO (H) SHOOGA 'ginger' (H) KAREE 'curry' (H)

(9) *cont.*

AMAKARAI 'sweet-pungent'			SATOO with SHOOYU 'soy sauce' or with SHOOGA 'ginger' etc. (Ki)
SHIOKARAI 'salt'	SHIO 'salt'	SHIO (strong)	UMI NO MIZU 'sea-water' (K)
SHOPPAI 'salty'	SHIO 'salt'	SHIO (strong)	
NIGAI 'bitter'	KUSURI 'medicine' (mentioned)	KINIINE 'quinine' KAFEIN 'caffeine'	I NO KUSURI 'stomach powders' (K)
HORONIGAI 'pleasantly bitter'		BIIRU 'beer'	
SHIBUI 'astringent'		SHIBUGAKI '*shibugaki*' (kind of persim-mon) CHA NO HA '(green) tea leaves'	JUKUSANAI KAKI 'unripe persimmons' (all) SHIBUGAKI (H)

Notable here is the prominence as exemplars of sugar (for AMAI 'sweet'), pickled plums (for SUPPAI 'sour'), red peppers (for KARAI 'pungent') and persimmons (either unripe or of the *shibugaki* variety) (for SHIBUI 'astringent'). All figure also in the informant's list, with the exception of red peppers (*toogarashi*); these are not generally employed in Japanese cuisine, and their omission by the informant simply reflects their lack of cultural salience in this respect. The major point of real disagreement concerns AMAKARAI 'sweet-pungent': as noted above, for the informant the application of this term is restricted to foods cooked with sugar and soy sauce, and she rejects the admissibility of other pungent substances such as ginger indicated in Kindaichi et al. (cf. Chapter 7, note 1).

From this initial characterization it seems clear that the denotation of the terms of AJI I takes in the parameters of gustation (i.e. the four common gustatory qualities of sweet, sour, salty and bitter) and, additionally, of pungency and astringency, which, as we have seen in 1.1, involve cutaneous and tactile sensations of the tongue and mouth. These are qualities which, we have suggested, are peculiarly unique to the workings of taste (1.3); interestingly, this range appears to be identical with that covered by the taste terms of Sanskrit as reported by Myers (cf. 1.2).

In this sub-section we also consider the question of the subjectivity of AJI I terms. The grammatical division between subjective and objective adjectives in Japanese was introduced in 4.2, and examined with respect to evaluative taste terms in 4.6. We have proposed (4.2) that descriptive taste terms also display subjective characteristics, a property which differentiates them linguistically from colour adjectives.

As we have seen, subjective adjectives are, first and foremost, directly predicable of first-person experiencers (cf. 4.6); *pace* Nishio (1972: 105), examples such as the following involving descriptive taste adjectives[9] are readily accepted by the informant:

(10) Kono mikan wa *suppaku* nai?
Watashi wa *amai* kedo.
THIS/MANDARIN/isolator/SOUR/negative, non-past
I/isolator/SWEET, non-past/BUT
'Aren't these mandarins sour?'
'I find them sweet.'

As noted in 5.2, -GARU derivatives of these terms also occur. Nishio (1972: 105) is reluctant to classify descriptive taste adjectives as subjective, and accepts only SUPPAGARU as a likely verbalization; however, we have seen that Martin's list (1975: 361–5) recognizes NIGAGARU and SHIBUGARU as potential formations (with 50–75 per cent acceptability ratings), and that derivatives such as KARAGARU are also readily accepted. Nevertheless, it is clear that judgements of acceptability vary in this area, and this may be due to several factors. For example, Martin points out (1975: 361) that a general finding of the reported survey was that -GARU derivatives from adjectives of lower frequency were less likely to be accepted than those from more common items. It seems probable that morphological factors are also at work here: thus the informant accepts AMAGARU and SHOPPAGARU as potential formations, but rejects *AMAZUPPAGARU, *AMAKARAGARU, *SHIOKARAGARU and *HORONIGAGARU, where the bases are clearly morphologically composite.[10] Restrictions of this kind are commonplace in the field of word-formation, and we may reasonably conclude that, in principle, descriptive taste adjectives with morphologically simple stems permit -GARU verbalization, although the resultant derivatives may differ in frequency of use.

Subjective adverbial uses of the kind noted for OISHII (4.6) do not appear to occur with descriptive taste adjectives. However, the use of derived evidential adjectives in -SOO-NA with reference to the outward manifestation of taste experience is unexceptional. Once again, adjectives such as SUPPAI 'sour' and NIGAI 'bitter' provide the most convincing examples, no doubt due to their ready association with clear external (facial) reactions:

(11) *suppasoo-ni* taberu 'eat in an apparently finding-sour manner'

(12) *nigasoo-ni* nomu 'swallow (e.g. medicine) in an apparently finding-bitter manner'

In the light of the above evidence, it is clear that descriptive taste adjectives in Japanese enter into subjective uses.[11] What is less clear is the generality of such uses, both in terms of their frequency and of the range of adjectives involved. To a greater extent than with evaluative taste terms, therefore, we must acknowledge the intermediate position of descriptive taste adjectives on the subjective/objective cline (cf. Chapter 4, note 16).

5.4.2 Intralingual relations

In this section we examine the intralingual meaning of the members of AJI I in terms of paradigmatic and syntagmatic relations.

5.4.2.1 Preliminaries

Three general issues which bear upon the semantic analysis in this section will be discussed briefly here: types of negation, the (paradigmatic) relation of incompatibility, and the notion of scalarity.

Firstly, two main types of negative responses, which we shall refer to as involving simple negation and WA-negation, occur in the data. They are illustrated in the following examples:

(B10)	Shibugaki wa *nigai*? SHIBUGAKI/isolator/BITTER, non-past 'Are *shibugaki* bitter?'	*Nigaku* nai. *Shibui.* BITTER/negative, non-past/ ASTRINGENT, non-past 'No, they're not. They're astringent.'
(B10')	Shibugaki wa *nigai*? 'Are *shibugaki* bitter?'	**Nigaku* wa nai kedo *shibui.* 'They're not *bitter*, but they're astringent.'
(B11)	Mikan wa *amai*? MANDARIN/isolator/ SWEET, non-past 'Are mandarins sweet?'	*Amaku* wa nai kedo *amazuppai.* SWEET/isolator/negative, non-past/BUT/SWEET-SOUR, non-past 'They're not *sweet*, but they're sweet-sour.'
(B11')	Mikan wa *amai*? 'Are tangerines sweet?'	**Amaku* nai. *Amazuppai.* 'No, they're not. They're sweet-sour.'

Here, *nigaku nai* (B10) is an example of simple negation, and *amaku wa nai* (B11) an example of WA-negation. Japanese *i*-adjectives form periphrastic negatives, consisting of the *-ku* form of the adjective (cf. 4.2) followed by forms of the negative item NAI: in formal terms, *nigaku nai* is thus the basic negative non-past form of NIGAI 'bitter', while *amaku wa nai* additionally contains the isolating particle WA interposed within the basic negative of AMAI 'sweet', *amaku nai.* This difference of negation type correlates in the examples with a difference in the overall pattern of the response: while the simple negation in (B10) is followed directly by an assertion, the WA-negation in (B11) is followed by KEDO 'but' plus a further assertion. These responses remain constant on different occasions of elicitation, and the alternative responses in (B10') and (B11') are consistently rejected by the informant.

This use of WA, and its semantic effects, have been little discussed in the literature (Kawakami, 1987: 58 is an exception). Its basic function is to mark the scope of the negation: in (B11), this is explicitly restricted to the applicability of the item AMAI 'sweet', this restriction in turn serving to signal the possibility that alternative, semantically neighbouring terms may indeed be applicable. Its overall effect here is thus one of attenuation; in English a similar effect is achieved by prosodic means, as indicated in the gloss, where italic represents the presence of a rise-fall tone. With antonyms, the two types of negation tend to exclude different sections of the associated scale: with reference to the pair ATSUI '(unpleasantly) hot'/SUZUSHII '(pleasantly) cool', Kawakami points out that *Kyoo wa atsuku nai* 'Today isn't hot' implicates *Kyoo wa suzushii* 'Today is cool', whereas *Kyoo wa atsuku wa nai* 'Today isn't *hot*' implicates rather *Kyoo wa maamaa suzushii* 'Today is reasonably cool' (1987: 58). Similarly, WA-negation was also required in the following example ((48) in 4.4) involving reference to the neutral region of the scale of palatability:

(13) *Mazuku* wa nai kedo taishite *umaku* mo nai.
'It doesn't taste bad but it doesn't taste very good, either.'

The occurrence of WA-negation (and the associated wider response pattern) in (B11) is thus suggestive of a closer semantic relationship between the terms AMAI 'sweet' and AMAZUPPAI 'sweet-sour' than between NIGAI 'bitter' and SHIBUI 'astringent'. Having introduced the distinction here, we shall explore the nature of this relationship in more detail later in this section.

The second question concerns the nature of the paradigmatic relation of incompatibility (cf 2.1.1). Lyons (1968: 458) defines this relation on the basis of contradictoriness between sentences and illustrates it by means of English colour terms. Thus, *Mary was wearing a red hat* implicitly denies *Mary was wearing a green (blue, white, yellow, etc.) hat* and vice versa, and the colour terms thereby constitute an incompatible set. (This definition of incompatibility also applies to opposites of various kinds; however, these are generally labelled in more specific terms (as complementaries, etc.: cf. 2.1.1) and they are not our concern here.) Another characteristic of colour terms, not drawn upon in this definition, is that two incompatible members of the set may not be applied simultaneously, in a coordinate construction, to the same object: sequences such as **Bill's car is red and it is white* or **Bill's red, white car* are unacceptable (as opposed to *Bill's car is red-and-white*, which is of course a different matter). It is perhaps this latter characteristic which relates most closely to the way in which the term 'incompatible' is used in everyday English and, as Lyons himself reports (1977: 288), there has been some confusion on this point in the linguistic literature.

It is instructive to consider how taste terms behave in this connection. Several pairs of AJI I terms are found to occur in parallel response patterns involving simple negation. Thus, in:

(B12) (=(B6))	Momo wa *suppai?* 'Are peaches sour?'	*Suppaku* nai. *Amazuppai.* 'No, they're not. They're sweet-sour.'
(B13)	Remon wa *amazuppai?* 'Are lemons sweet-sour?'	*Amazuppaku* nai. *Suppai.* 'No, they're not. They're sour.'

it is shown that the pair of terms SUPPAI 'sour' and AMAZUPPAI 'sweet-sour' can each be asserted after the simple negation of the other. We shall say that SUPPAI and AMAZUPPAI figure in bilateral simple negation: that is to say, we can deny, without qualification, that a thing is SUPPAI 'sour' and simultaneously assert that it is AMAZUPPAI 'sweet-sour', and vice versa. Now the colour terms discussed above, such as RED, GREEN, etc., negate each other in the same manner (*Is grass red? No, it's green,* etc.). However, with respect to the second property raised above, i.e. the impossibility of simultaneous application, we find that this holds for some, but not all, of the AJI I terms involved: for example, it holds for the two terms SUPPAI 'sour' and AMAI 'sweet' (we cannot say that something is **suppakute amai* 'sour and sweet'), but it does not hold for the terms NIGAI 'bitter' and SHIBUI 'astringent', which, although figuring in bilateral simple negation in exactly the same way as SUPPAI and AMAI, can, unlike them, be predicated simultaneously of the same subject (e.g. *Desugita ocha wa nigakute shibui* 'Over-infused tea is bitter and astringent'). Similar restrictions appear to apply with the corresponding English terms: we cannot naturally speak of **a sweet, sour drink* (*a sweet-and-sour drink,* or *a sweet-sour drink,* is again a different matter), whereas to speak of *a bitter, astringent drink* seems unobjectionable.

There are, then, two separate properties at issue here, and our description will distinguish terminologically between them. The basic notion involved in the paradigmatic relation of incompatibility is that of difference within similarity (cf. Lyons, 1968: 459). In the literature of anthropological linguistics (e.g. Frake, 1961), the term 'contrast' ('levels of contrast', 'contrast set',[12] etc.) has been applied to this type of relation, and we propose to adopt this usage in our analysis. As far as the definition of the relation is concerned, a criterion of the following kind based on negation appears to be appropriate: two terms are related by contrast if it is possible simultaneously to negate the one and assert the other. Thus, from the acceptability of *Blood is not green, it's red* and *Grass is not red, it's green* we may conclude that RED and GREEN are contrasting terms. Similarly, from their occurrence in bilateral simple negation noted above, we may conclude that SUPPAI 'sour' and AMAI 'sweet' are related by contrast, as are NIGAI 'bitter' and SHIBUI 'astringent'. For the second property we are now free to speak of syntagmatic incompatibility (relative to a given construction) (cf. 2.1.2). Thus, while the pairs RED/GREEN, SUPPAI/AMAI and NIGAI/SHIBUI are all pairs of (paradigmatically) contrasting terms, the first two pairs are, in addition, (syntagmatically) incompatible in coordinate constructions, unlike NIGAI and SHIBUI. The fact that all contrasting colour terms are syntagmatically incompatible in such cases, but that all taste terms are not, obviously constitutes an important semantic difference between the structure of the two fields.

Finally, in the course of our analysis we will find some items which, not being antonyms, appear to be semantically related as terms on a scale. An example is the pair SHIOKARAI 'salt' and SHOPPAI 'salty' which, intuitively, range over a single scale ('saltiness'), with SHIOKARAI occupying a more extreme position than SHOPPAI. The question arises whether these cases should, or can, be handled by means of a special paradigmatic scalar relation and, if so, how this relation is to be defined. Examples of this kind have been discussed by various linguists (e.g. Lehrer and Lehrer, 1982), and the whole area has been critically surveyed by Westney (1986). His conclusion is that, as a proposed intralingual semantic relation between lexical items, scalarity is beset by theoretical difficulties and that few sets of items constitute scales in any well-defined sense (1986: 351–2). Accordingly, we shall not appeal to such a notion in our analysis. While pointing out that the items concerned appear to denote similar qualities which differ in intensity, we shall analyse them intralingually as being related by contrast or by hyponymy, according to the evidence in each case.

5.4.2.2 Response patterns: main terms

In the remaining sub-sections we present the analysis of AJI I terms in progressive fashion, the results of earlier stages being refined, and in some cases considerably modified, in succeeding stages.

Our initial aim is to obtain a rough picture of the structure of the system by surveying response patterns to question-frame (B). We begin here with non-MI GA ARU terms, which we may refer to as 'main' terms; -MI GA ARU terms in some ways stand apart, and are dealt with as a group in the next sub-section (5.4.2.3).

(B12) and (B13) have shown that SUPPAI 'sour' and AMAZUPPAI 'sweet-sour' constitute a pair of contrasting terms, figuring in bilateral simple negation; this relation is further reflected in the following description of pomegranates (*zakuro*), attested in conversation:

(14) *Suppai* no to *amazuppai* no ga atte...
'There are sour ones and sweet-sour ones...'

The following examples show that SUPPAI 'sour' and AMAI 'sweet' also constitute a contrast pair:

(B14)	Banana wa *suppai*? 'Are bananas sour?'	*Suppaku* nai. Banana wa *amai*. 'No, they're not. Bananas are sweet.'
(B15)	Remon wa *amai*? 'Are lemons sweet?'	*Amaku* nai. Remon wa *suppai*. 'No, they're not. Lemons are sour.'

However, as already foreshadowed, the third pairing of these three terms, AMAI 'sweet' and AMAZUPPAI 'sweet-sour', behave differently:

(B16)	Banana wa *amazuppai*? 'Are bananas sweet-sour?'	*Amazuppaku* nai. *Amai*. 'No, they're not. They're sweet.'
(B17) (=(B11))	Mikan wa *amai*? 'Are mandarins sweet?'	*Amaku* wa nai kedo *amazuppai*. 'They're not *sweet*, but they're sweet-sour.'
(B17') (=(B11'))	Mikan wa *amai*? 'Are mandarins sweet?'	**Amaku* nai. *Amazuppai*. 'No, they're not. They're sweet-sour.'

The response to (B17) exhibits WA-negation, and simple negation (B17') is rejected. At this stage, the semantic relationship between AMAZUPPAI 'sweet-sour' and AMAI 'sweet' is therefore assumed to be different in some way from that between the other pairs of terms.

As illustrated by these examples, the procedure employed here is to use question-frame (B) (*X wa Y?* 'Is X Y?') to elicit response patterns for permutations of AJI I expressions. In each case, the precise formulation of the question used is based on the information gained from the earlier responses to question-frame (A) (*X wa donna aji ga suru?* 'What kind of a taste does X have?'). For example, given the information from (A) that lemons are said to be SUPPAI 'sour' ((A5)) and that onions are said to be KARAI 'pungent' ((A8)), we can devise the following (B)-type questions to test the response patterns for the two terms SUPPAI and KARAI:

(B18)	Remon wa *karai*? 'Are lemons pungent?'	*Karaku* nai. *Suppai.* 'No, they're not. They're sour.'
(B19)	Tamanegi wa *suppai*? 'Are onions sour?'	Uun, *suppaku* nai. Tamanegi wa *karai.*[13] 'No, they're not. Onions are pungent.'

On the basis of such response patterns, each pair of AJI I terms can initially be classified as either figuring in bilateral simple negation or not. Thus, (B12)–(B19) yield the pairs SUPPAI 'sour'/AMAZUPPAI 'sweet-sour', SUPPAI/AMAI 'sweet' and SUPPAI/KARAI 'pungent' as falling within the bilateral pattern, and the pair AMAI/AMAZUPPAI as standing outside it.

All permutations of pairs of main AJI I terms were tested in this way. Bilateral simple negation was found to hold for all pairs except the following:

AMAI 'sweet'/AMAZUPPAI 'sweet-sour'
AMAI 'sweet'/AMAKARAI 'sweet-pungent'
NIGAI 'bitter'/HORONIGAI 'pleasantly bitter'
KARAI 'pungent'/SHIOKARAI 'salt'
SHOPPAI 'salty'/SHIOKARAI 'salt'
KARAI 'pungent'/PIRITTO 'piquantly'

The relevant data for the above non-bilateral pairs is as follows. First, for AMAI 'sweet' and AMAKARAI 'sweet-pungent':

(B20)	Mame no kanroni wa *amakarai*? 'Are sweet-boiled beans sweet-pungent?'	*Amakaraku* nai. *Amai.* 'No, they're not. They're sweet.'
(B21)	Sukiyaki wa *amai*? 'Is sukiyaki sweet?'	Soo-ne, *amaku* wa nai kedo *amakarai*. 'Well, it's not *sweet*, but it's sweet-pungent.'

As shown, simple negation holds in one direction ((B20)), with WA-negation occurring in the other. AMAKARAI thus resembles AMAZUPPAI in that both items enter into bilateral simple negation with one of the AJI I terms that figures in their morphological make-up (KARAI 'pungent' and SUPPAI 'sour', respectively), but not with the other (AMAI 'sweet').

The same pattern holds for NIGAI 'bitter' and HORONIGAI 'pleasantly bitter':

(B22)	Konagusuri wa *horonigai*? 'Are medical powders pleasantly bitter?'	Uun, *horonigaku* nai. *Nigai.* 'No, they're not. They're bitter.'
(B23)	Chokoreeto no burakku wa *nigai*? 'Is plain chocolate bitter?'	*Nigaku* wa nai kedo *horonigai*. 'It's not *bitter*, but it's pleasantly bitter.'

With KARAI 'pungent' and SHIOKARAI 'salt', new patterns are found:

(B24)	Shooyu wa *shiokarai*? 'Is soy sauce salt?'	Uun, *shiokaraku* nai. Shooyu wa *karai* dake. NO/SALT/negative, non-past/SOY SAUCE/ isolator/ PUNGENT, non-past/ONLY 'No, it isn't. Soy sauce is just pungent.'

(B25) Umi no mizu wa *karai*? — Un, *karai*. *Shiokarai.*
'Is sea-water pungent?' — 'Yes, it is. It's salt.'

With this pair, there is simple negation in one direction ((B24), although with the addition of DAKE 'only'); in the other direction, however, there is no negation at all, but affirmation. The examples strongly suggest a relation of hyponymy, with KARAI 'pungent' functioning as the superordinate of SHIOKARAI 'salt'.

The pair SHIOKARAI 'salt' and SHOPPAI 'salty' also show a new pattern:

(B26) Umi no mizu wa *shoppai*? — *Shoppai* dokoroka *shiokarai.*
'Is sea-water salty?' — 'Far from being (just) salty, it's salt.'

(B27) Kobucha wa *shiokarai*? — Kobucha wa *shiokaraku* nai. Chotto *shoppai* dake.
'Is *kobucha* salt?' — 'No, it isn't. It's just somewhat salty.'

The construction with DOKOROKA 'far from' in (B26) expresses a strong denial of the preceding predication; commonly, although by no means exclusively, it occurs where a suggestion is overruled as being insufficiently extreme (cf. *Indo wa attakai-daroo?* 'India is warm, I suppose' *Attakai dokoroka atsui yo* 'Warm? It's hot'). Its replacement here by simple negation is rejected by the informant, and the examples suggest that SHIOKARAI and SHOPPAI stand in a scalar relationship, as discussed in 5.4.2.1. However, we saw there that, as a paradigmatic relation, scalarity lacks a firm theoretical basis. Accordingly, we analyse SHIOKARAI and SHOPPAI intralingually as contrast terms, and interpret the occurrence in one direction of DOKOROKA, rather than simple negation, as reflecting the evident close extralingual relationship, where both terms denote the quality of saltiness, but in differing intensities. Cf. also the following example from the data:

(B28) Shiomizu wa *shiokarai*? — Soo-ne, ammari koi to ne. Usukereba *shoppai* gurai da kedo.
WELL/EXCEEDINGLY/ CONCENTRATED, non-past/IF/tag/DILUTE, provisional/SALTY,

	non-past/EXTENT/copula, non-past/BUT
'Is a salt-and-water solution salt?'	'Well, it is if it's very concentrated. If it's weak, it's only salty, though.'

where the proposed difference of intensity is clearly correlated with the dimension USUI 'weak, dilute'/KOI 'strong, concentrated'. The following example is also relevant:

(B29)	Shiomizu wa *karai*?	Un, chotto koi to *karai*.
		YES/SOMEWHAT/ CONCENTRATED, non-past/IF/ PUNGENT, non-past
	'Is a salt-and-water solution pungent?'	'Yes, if it's somewhat concentrated.'

since, in conjunction with (B28), it clearly supports the proposed hyponymy of SHIOKARAI with KARAI suggested earlier.

Apropos of these three terms, we should note that this is an area of considerable dialectal complexity. In answer to the question concerning the taste of salt in the Linguistic Atlas of Japan, the main responses were KARAI, SHOPPAI and SHIOKARAI, with KARAI and SHOPPAI prominent in western and eastern Japan respectively, and SHIOKARAI scattered throughout the country (Kokuritsu Kokugo Kenkyuujo, 1966b: 24). Importantly, however, it is clear that all three terms are used in Tokyo, where SHOPPAI is apparently a relatively recent arrival: thus Yamada, a Tokyo speaker, includes SHOPPAI as a basic member of his active vocabulary while noting its instability as a competitor of KARAI and SHIOKARAI (1972: 31, 36), and Fujisawa notes the use of SHOPPAI among young Tokyo speakers (1975: 54). Some Linguistic Atlas respondents indicated stylistic distinctions, characterizing KARAI as more elegant than SHOPPAI, and many dictionaries comment on the informal/colloquial status of the latter (e.g. Hayashi et al., 1987: 460). As far as the present analysis is concerned, the key fact remains that, as is clear from the data, the informant uses all three terms in identical stylistic circumstances (i.e. informal (adult) speech), with the semantic distinctions of intensity as indicated. Thus in generic contexts the informant reserves SHIOKARAI[14] for the taste of substances such as sea-water which are not normally consumed and applies SHOPPAI to ordinary items of diet (such as *kobucha* tea); for salt-and-water solutions, she uses both according to strength. Similarly, SHOPPAI is not permitted for the informant in the following specific statement:

(15) Kore wa shio ga kikisugite *karai/shiokarai/*shoppai.*
'This has too much salt in it and tastes pungent/salt/*salty.'

SHOPPAI and SHIOKARAI are likewise not reversible in the constructions with DOKOROKA and DAKE in (B26) and (B27) above and, as we shall see in 5.4.2.5, the terms differ in affective value. In general, dictionaries fail to distinguish semantically between this pair; the work of Hayashi et al., however, is an exception, and their definition of SHIOKARAI (*Embun ga, shita o sasu yoo ni kanjirareru hodo, kitsui* 'Salt content is so intense as to be felt as stinging the tongue' (1987: 383)) clearly points to excessive intensity.[15]

Before considering the response patterns for the pair KARAI 'pungent' and PIRITTO 'piquantly', we must first comment further on the lexical status of this latter item. We have seen that, in the response to (A10), PIRITTO occurs as an adverb modifying KARAI. KARAI is in fact the only lexical item with which PIRITTO collocates in this construction: that is to say, in modifier-head constructions, there is a relationship of syntagmatic implication between these terms.[16] In addition to (or, in certain cases, instead of) functioning as adverbs, many mimetic items also form derived verbs ending in -SURU: PIRITTO is such an item, forming a derived stative verb with the basic forms *piritto-shite iru* (predicative) (generally contracted to *piritto-shiteru* in informal speech) and *piritto-shita* (adnominal). This derived item (i.e. the verb PIRITTO-SHITE IRU) occurs in (B)-type questions and in some responses:

(A10)	Karashina wa donna aji ga suru? 'What kind of a taste does mustard-plant have?'	Karashina wa *piritto karai*/ *Karashina wa *piritto-shiteru.* 'It's piquantly pungent.'
(B30)	Karashina wa *karai*? 'Is mustard-plant pungent?'	Un, chotto *piritto karai* ne./ Un, chotto *piritto-shiteru* ne./Un, chotto *karai*. Chotto *piritto-shiteru.* 'Yes, it's somewhat piquantly pungent.'/'Yes,

		it's somewhat piquant.'/ 'Yes, it's somewhat pungent. It's somewhat piquant.'
(B31)	Karashina wa *piritto-shiteru*?/ *Karashina wa *piritto karai*? 'Is mustard-plant piquant?'	Un, *piritto karai* ne./Un, chotto *piritto-shiteru* ne. 'Yes, it's piquantly pungent.'/ 'Yes, it is somewhat.'
(B32)	Umi no mizu wa *piritto-shiteru*? 'Is sea-water piquant?'	Uun, *piritto-shitenai* yo. *Shiokarai* yo./ *Uun, *piritto karaku* nai yo. *Shiokarai* yo. 'No, it isn't. It's salt.'

As illustrated in these question-and-answer pairs from the data, the main pattern appears to be that PIRITTO-SHITE IRU is preferred in questions and in negative responses (cf. (B31), (B32)). In positive replies, the informant prefers PIRITTO (+KARAI) in response to question-frame (A) (cf. (A10)), while both may occur in response to (B)-type questions (cf. (B31)). The question arises whether, in the cases where both the adverb and the derived verb appear, there is any associated semantic difference. For the informant, the two types of response, as in the following examples from (B30), are tantamount to repetitions of each other:

(16) Un, chotto *piritto karai* ne.
'Yes, it's somewhat piquantly pungent.'

(17) Un, chotto *piritto-shiteru* ne.
'Yes, it's somewhat piquant.'

and the following sentence, asserting PIRITTO (+KARAI) and denying PIRITTO-SHITE-IRU, is regarded as contradictory:

(18) **Piritto karai* kedo *piritto-shitenai*.
'It's piquantly pungent but it's not piquant.'

We thus conclude that the two items are synonymous. From this point on we shall refer to them in tables, etc. by the conflated label PIRITTO(-SHITE IRU) 'piquant', it being understood that their distribution is determined by the factors mentioned above.

We may now return to the patterns which hold with (B)-type questions between this item and KARAI. The pattern in one direction has already been illustrated in (B30): as with the pair SHIOKARAI 'salt' and KARAI in (B25), there is no negation, but affirmation. The pattern in the opposite direction is shown by:

(B33)	Shooyu wa *piritto-shiteru*?	Uun, *piritto-shitenai.* Shooyu wa *karai* dake.
	'Is soy sauce piquant?'	'No, it isn't. Soy sauce is just pungent.'

and

(B34)	Shooyu wa *piritto-shiteru*?	Uun, shooyu wa *piritto-shita karasa* janai. NO/SOY SAUCE/isolator/ PIQUANT, past/ PUNGENCY/copula, negative, non-past
	'Is soy sauce piquant?'	'No, it isn't a piquant pungency in the case of soy sauce.'

Here (B33), with DAKE 'only', is again parallel to the case of SHIOKARAI/ KARAI (B24), and, together with the earlier examples, likewise suggests a relationship of hyponymy, with KARAI 'pungent' functioning as the superordinate of PIRITTO-SHITE IRU 'piquant'. This is further supported by the response in (B34), with the adnominal form *piritto-shita* modifying the noun KARASA 'pungency'.

The diagram on the following page summarizes the disposition of the items in terms of the response patterns discussed in this sub-section:

(19)

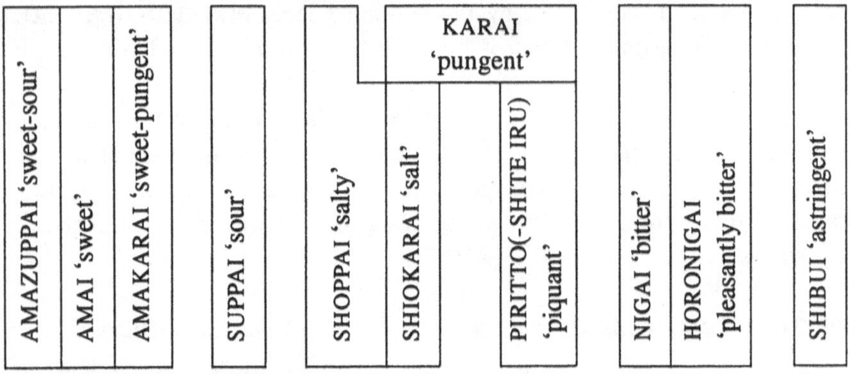

The items are grouped in five boxes, as shown. Terms in different boxes (e.g. AMAI and SUPPAI) enter into bilateral simple negation and are clearly related by contrast; the same applies to non-adjoining terms within the same box (e.g. AMAZUPPAI and AMAKARAI). Adjoining terms in the same box separated by a vertical line have simple negation in one direction and WA-negation (e.g. AMAZUPPAI and AMAI) or denial by DOKOROKA (SHIOKARAI and SHOPPAI) in the other; these terms thus appear to be particularly close in meaning, and outstanding details will be examined further below. Finally, adjoining terms separated by a horizontal line (e.g. KARAI and SHIOKARAI) stand in a relationship of hyponymy, the upper term being the superordinate. The semantic analysis of main AJI I terms is further developed in later sub-sections.

5.4.2.3 Response patterns: -MI GA ARU terms

In this sub-section we turn our attention to the -MI GA ARU terms of AJI I. We begin by considering the semantic disposition of individual terms vis-a-vis their 'corresponding' main AJI I terms. Thus, of the five terms in the system, AMAMI GA ARU 'has a sweetness', NIGAMI GA ARU 'has a bitterness' and SHIBUMI GA ARU 'has an astringency' may be expected, assuming a straightforward morphological-semantic correlation, to relate semantically to AMAI 'sweet', NIGAI 'bitter' and SHIBUI 'astringent', respectively. The response pattern in one direction is illustrated below:

(B35)	Ninjin wa *amai*?	Soo-ne, chotto *amami ga aru*.
	'Are carrots sweet?'	'Well yes, they have a slight sweetness.'

(B9) Sakkarin wa *nigai*? — Soo-ne, chotto *nigami ga aru* ne.

'Is saccharine bitter?' — 'Well yes, it has a slight bitterness.'

(B36) Budoo no kawa wa *shibui*? — Soo-ne, chotto *shibumi ga aru*.

'Are grape skins astringent?' — 'Well yes, they have a slight astringency.'

The pattern here may be characterized basically as assertion, and simple negation is impossible. Patterns in the reverse direction are similar:

(B37) Banana wa *amami ga aru*? — Soo-ne, kanari *amami* ga tsuyoi.
WELL/ CONSIDERABLY/ SWEETNESS/subject/ STRONG, non-past

'Do bananas have a sweetness?' — 'Well yes, they have a pretty strong sweetness.'

(B38) Biiru wa *nigami ga aru*? — Soo-ne, chotto *nigai*.

'Does beer have a bitterness?' — 'Well yes, it's somewhat bitter.'

(B39) Shibugaki wa *shibumi ga aru*? — Soo-ne, kanari *shibumi ga aru* ne. *Shibui*.

'Do *shibugaki* have an astringency?' — 'Well yes, they have a considerable astringency. They're astringent.'

As we have seen in 5.2, the other -MI GA ARU terms, SAMMI GA ARU 'has a sourness' and SUIMI GA ARU 'has a tartness', are not morphologically related to any main terms. However, their derivation from the noun SAN 'acid' and the adjective SUI 'sour', respectively, points to SUPPAI 'sour' as a likely main term counterpart. With SAMMI GA ARU and SUPPAI, assertive patterns indeed occur in both directions (although WA-negation has also been seen in (B5), repeated below):

(B40)	Moka wa *suppai*? 'Is mocha sour?'	Soo-ne, chotto *sammi ga aru.* 'Well yes, it has a slight sourness.'
(B5)	Koogyoku wa *suppai*? 'Are Jonathans sour?'	*Suppaku* wa nai kedo *sammi ga aru.* 'They're not *sour*, but they do have a sourness.'
(B41)	Remon wa *sammi ga aru*? 'Do lemons have a sourness?'	Soo yo. 'Certainly.'

With SUIMI GA ARU and SUPPAI, however, we find quite different patterns, namely simple negation in both directions:

(B42)	Momo wa *suppai*? 'Are peaches sour?'	*Suppaku* nai kedo *suimi wa aru.* 'No, they're not, but they do have a tartness.'
(B43)	Remon wa *suimi ga aru*? 'Do lemons have a tartness?'	Remon? *Suimi nai.* Remon wa *suppai.* 'Lemons? No, they don't. Lemons are sour.'

As we shall see, the semantics of SUIMI GA ARU is complicated by its special relationship with AMAI 'sweet', which disturbs the expected correlation with SUPPAI here. At the same time we may note that the form of the response in (B42) is nevertheless suggestive of some semantic connection between the two terms: while showing the simple negation of SUPPAI, it also contains KEDO 'but' and the isolating particle WA, which here carries a concessive force. These points will be taken up further below.

Let us now consider more closely how -MI GA ARU terms as a group differ from main terms. Central here are different response patterns for parallel questions involving main terms and the 'corresponding' -MI GA ARU terms in cases like the following:

(B44) Koohii wa *amai*? Uun, *amaku* nai. Koohii wa chotto *nigai.*
'Is coffee sweet?' 'No, it isn't. Coffee is somewhat bitter.'

Here we see the expected response, with the simple negation of one main term (AMAI) followed by the assertion of another, contrasting term (NIGAI). The response to a parallel question with the corresponding -MI GA ARU term, however, is as follows:

(B45) Koohii wa *amami ga aru*? Uun, *amami ga nai.* Koohii wa *nigai* dake.
'Does coffee have a sweetness?' 'No, it doesn't. Coffee is just bitter.'

with DAKE 'only, just' in the answer. Whereas, in (B44), the plain assertion of NIGAI 'bitter' is compatible with the denial of AMAI 'sweet', (B45) indicates that the assertion of NIGAI alone is not sufficient to justify the denial of AMAMI GA ARU 'has a sweetness' but must be followed by DAKE, which explicitly excludes the possibility of a taste combination. Similar examples occur with the other -MI GA ARU terms:

(B46) Remon wa *nigami ga aru*? Uun, *nigami ga nai.* Remon wa *suppai* dake.
'Do lemons have a bitterness?' 'No, they don't. Lemons are just sour.'

(B47) Natsumikan wa *shibumi ga aru*? Uun, natsumikan wa *shibumi ga nai*. Are wa *suppai* dake.
'Do *natsumikan* (a type of orange) have an astringency?' 'No, they don't. They're just sour.'

(B48) Shibugaki wa *sammi ga aru*? Uun, *sammi ga nai.* Shibugaki wa *shibui* dake.
'Do *shibugaki* have a sourness?' 'No, they don't. *Shibugaki* are just astringent.'

(B49) Meron wa *suimi ga aru*? — Uun, *suimi ga nai.* Meron wa *amai* dake.

'Do melons have a tartness?' — 'No, they don't. Melons are just sweet.'

This is a clear and consistent difference in patterning, and it leads to the hypothesis that, unlike their corresponding main terms, -MI GA ARU terms predicate a given quality as a tinge, rather than as a predominant component, of an overall taste: tinges being readily compatible with contrasting predominant tastes, there is a need explicitly to rule out such combinations in examples such as those above.

The question of taste combinations is clearly of major importance here. While the combinatory possibilities of AJI I terms will be examined in detail in the next sub-section, it is appropriate to introduce some general considerations on the topic at this point.

Taking (English) colour terms as a point of comparison, we note two broad patterns of combining contrasting terms in that domain, correlating with clear psychophysical differences. First, contrast terms may be combined in coordinate expressions involving the conjunction AND: just as an object may be described as RED, so objects may be described as RED AND BLACK, as RED AND YELLOW, as RED, WHITE AND BLUE, and so on. Such combinations are unrestricted, and the objects they describe are multi-coloured. The second pattern involves expressions such as GREY-GREEN, GREENISH-BLUE, BLUEY-GREY, ORANGEY-YELLOW, etc. These expressions typically contain two terms, the first, which is frequently a derivative, generally serving to modify the second. Moreover, certain restrictions based on the semantic relations that hold between colour terms come into play here, so that combinations such as *YELLOW-BLUE, *BLACKISH-WHITE, etc. are normally unacceptable. Rather than multi-colouring, where we have a juxtaposition of two or more discrete colours, these expressions denote a colour blend or shade falling between two basic colours (cf. Kay and McDaniel, 1978: 622–3). Such blends may involve both basic colours in equal weights, or one predominant colour tinged with the other, subsidiary colour. Note that derivatives used alone (*bluish leaves, orangey bricks*) also suggest intermediate shades, with the remaining member unspecified. With colour then, the psychophysical distinction between multi-colouring and colour blends is basically clear and, in English at least, it is reflected in a clear difference of linguistic formulation.

By contrast, the psychophysical situation for taste is much less clear-cut, and the status of a comparable dichotomy between multi-taste combinations and taste blends is problematical: is the taste of a sweet-and-sour sauce, for example, perceived as a conjunction of these two discrete qualities or as a single blended quality? Our observations for taste must accordingly centre on the linguistic data.

That some AJI I terms may combine in coordinate structures is already clear from the following responses to question-frame (A):

(A17)	Momo wa donna aji ga suru? 'What kind of a taste do peaches have?'	*Amakute suimi ga aru.* 'They're sweet and have a tartness.'
(A19)	Umi no mizu wa donna aji ga suru? 'What kind of a taste does sea-water have?'	*Karakute* chotto *nigami ga aru.* 'It's pungent and has a slight bitterness.'
(A20)	Desugita ocha wa donna aji ga suru? 'What kind of a taste does over-infused green tea have?'	*Nigakute shibumi ga aru.* 'It's bitter and has an astringency.'

Here we see the standard coordinating construction for predicates, involving the conjunctive form of non-final predicators. All responses involve two-term combinations, with, in each case, a -MI GA ARU term as one of the members. The following responses have also been seen:

(A27)	Tomato wa donna aji ga suru? 'What kind of a taste do tomatoes have?'	*Sammi ga atte* chotto *aokusai.* 'They have a sourness and are somewhat grassy-flavoured.'
(A18)	Sakkarin wa donna aji ga suru? 'What kind of a taste does saccharine have?'	*Amai* kedo *nigami ga aru.* 'It's sweet but has a bitterness.'

where again we find -MI GA ARU terms predicated together with other taste terms: with an AJI III term (AOKUSAI 'grassy-flavoured') in (A27), and in an adversative construction with a main AJI I term (AMAI 'sweet') in (A18). In contexts of the present type, it seems clear that -MI GA ARU terms are commonly predicated as an associated quality rather than as an independent one.

Apparent counterexamples may in fact involve implicit combinations of terms. Thus, in the following example we find AMAMI GA ARU used alone:

(A32) Sukotchiuisukii wa donna aji ga suru? — Chotto *amami ga aru.*
'What kind of a taste does Scotch whisky have?' — 'It has a slight sweetness.'

However, we need to take into account here the semantic status of SUKOTCHIUISUKII as a hyponym of UISUKII 'whisky', and the fact that a question about the taste of whisky in general is answered as follows:

(A33) Uisukii wa donna aji ga suru? — Chotto *karai.*
'What kind of a taste does whisky have?' — 'It's somewhat pungent.'

The response to (A32) is thus to be set against this general taste norm for whisky, and the term KARAI 'pungent' may be regarded as implicitly present. The following example specifies an overall taste, although there is no main taste component to be described by a single taste term:

(A29) Ninjin wa donna aji ga suru? — Maa, ninjin no aji da kedo. Chotto *amami ga aru.*
'What kind of a taste do carrots have?' — 'Well, they taste like carrots – they have a slight sweetness.'

Though not seen in the data so far, we also find coordination between main AJI I terms (cf. 5.4.2.1). Thus, in response to the (A)-type question relating to overinfused tea, both (A20) and (A34) were elicited:

(A20) Desugita ocha wa donna aji ga suru? — *Nigakute shibumi ga aru.*
'What kind of a taste does over-infused green-tea have?' — 'It's bitter and has an astringency.'

(A34)	Desugita ocha wa donna aji ga suru?	*Nigakute shibui.*
	'What kind of a taste does over-infused green-tea have?'	'It's bitter and astringent.'

According to the informant, the response in (A20) carries the implication that the predominant taste is bitter (NIGAI), whereas in (A34) both bitter and astringent (SHIBUI) tastes are presented on an equal footing.

In taste, then, we find combinations of main terms as well as combinations of a main term with what we have called a 'tinge' term, and tinge terms are commonly predicated as (subsidiary) members of such combinations. There are perhaps some echoes here of the two types of combination noted for English colour terms. Like GREENISH, BLUEY, etc., -MI GA ARU terms are morphological derivatives from their corresponding main terms, and they may be used in a similar way as modifiers: thus, just as we may speak of *a yellowish brown* but not of **a yellow brown*, in Japanese we may speak of *shibumi ga aru nigasa* 'a bitterness which has an astringency' but not of **shibui nigasa* 'an astringent bitterness'.

Examples given above are also of help in considering further the meaning of SUIMI GA ARU. In particular (B49), repeated below, contrasts clearly with (B43) and (B50):

(B49)	Meron wa *suimi ga aru*?	Uun, *suimi ga nai.* Meron wa *amai* dake.
	'Do melons have a tartness?'	'No, they don't. Melons are just sweet.'
(B43)	Remon wa *suimi ga aru*?	Remon? *Suimi nai.* Remon wa *suppai.*
	'Do lemons have a tartness?'	'Lemons? No, they don't. Lemons are sour.'
(B50)	Shibugaki wa *suimi ga aru*?	Nai. Shibugaki wa *shibui.*
	'Do *shibugaki* have a tartness?'	'No, they don't. *Shibugaki* are astringent.'

Unlike (B49), neither of these responses contain DAKE and, taken together, the examples suggest that SUIMI GA ARU is compatible with AMAI 'sweet', but not, for example, with SUPPAI 'sour' or SHIBUI 'astringent'. Indeed, the co-occurrence of AMAI with SUIMI GA ARU in the data is striking; cf. (A17) and (B4):

(A17)	Momo wa donna aji ga suru? 'What kind of a taste do peaches have?'	*Amakute suimi ga aru.* 'They're sweet and have a tartness.'
(B4)	Momo wa *amai*? 'Are peaches sweet?'	*Amai* kedo *suimi mo aru.* 'Yes, they are, but they also have a tartness.'

Testing of combinations with the informant in the next sub-section will confirm that SUIMI GA ARU is in fact compatible only with AMAI: that is, *amakute suimi ga aru* 'is sweet and has a tartness' occurs, but comparable collocations with other main terms are rejected, as are sentences of the form **X wa suimi ga aru kedo amaku nai* 'X has a tartness but is not sweet'. As a tinge term, SUIMI GA ARU is thus exceptional in that it carries a definite implication as to the predominant taste quality involved: this must be sweet (AMAI).

Returning to the status of SUIMI GA ARU in terms of its correspondence to a particular main term, let us look again at (B42):

(B42)	Momo wa *suppai*? 'Are peaches sour?'	*Suppaku* nai kedo *suimi wa aru.* 'No, they're not, but they do have a tartness.'

As noted earlier, the response shows the simple negation of SUPPAI; this is now to be expected, given the relation just referred to between SUIMI GA ARU and AMAI, and the relation of contrast between AMAI and SUPPAI. At the same time, however, we saw that the form of the response in (B42) is also suggestive of a certain semantic connection: we propose that this connection is extralingual, and that SUIMI GA ARU denotes a sour tinge, within a predominantly sweet taste.[17] As we know, the term AMAZUPPAI 'sweet-sour' is also applied to taste qualities which combine sweetness and sourness, and our data to this point are clearly suggestive of semantic overlap with (AMAI +) SUIMI GA ARU. The central case here is that of peaches (*momo*). In the discussion of extralingual meaning (5.4.1), it was noted that the informant listed this fruit under taste norms for both these terms, and responses to questions confirm that both may be predicated of MOMO; the relevant examples are repeated here:

(A3)	Momo wa donna aji ga suru? 'What kind of a taste do peaches have?	*Amazuppai.* 'They're sweet-sour.'
(A17)	Momo wa donna aji ga suru? 'What kind of a taste do peaches have?'	*Amakute suimi ga aru.* 'They're sweet and have a tartness.'
(B4)	Momo wa *amai*? 'Are peaches sweet?'	*Amai* kedo *suimi mo aru.* 'Yes, they are, but they also have a tartness.'
(B6)	Momo wa *suppai*? 'Are peaches sour?'	*Suppaku* nai. *Amazuppai.* 'No, they're not. They're sweet-sour.'
(B42)	Momo wa *suppai*? 'Are peaches sour?'	*Suppaku* nai kedo *suimi wa aru.* 'No, they're not, but they do have a tartness.'

At the same time, the data suggest that other fruits, such as mandarins (*mikan*), may be described as AMAZUPPAI but not readily as SUIMI GA ARU:

(A4)	Mikan wa donna aji ga suru? 'What kind of a taste do mandarins have?'	*Amazuppai.* 'They're sweet-sour.'
(B11)	Mikan wa *amai*? 'Are mandarins sweet?'	*Amaku* wa nai kedo *amazuppai.* 'They're not *sweet*, but they're sweet-sour.'

Cf. also:

(B51)	Mikan wa *suppai*? 'Are mandarins sour?'	*Suppaku* nai. *Amazuppai.* 'No, they're not. They're sweet-sour.'

Responses with SUIMI GA ARU are not readily forthcoming here: for the informant, the application of this term centres on peaches, while AMAZUPPAI takes in a wider range of fruits, as well as cooked dishes such as sweet-and-sour pork. Additional examples of the use of SUIMI GA ARU by the informant are attested in the secondary conversational data, firstly with reference to the taste of mangoes:

(20) *Amai* bakari de *suimi ga nai*-deshoo.
'They (= mangoes) are just sweet and have no tartness, you see.'

Mangoes are an exotic fruit for the Japanese; despite their physical resemblance to peaches, in taste they lack tartness. The second example occurs in a comment on the suitability of specific apricots – another peach-like fruit – for jam-making:

(21) *Suimi ga nakute* oishiku nai no ne, *amai* bakkashi de...Anzu tte *amazuppai* n janai? *Amazuppaku* nai to zenzen oishiku nai.
'The thing is they (= these apricots) have no tartness and don't taste good, being just sweet...Apricots should be sweet-sour, right? If they're not sweet-sour, they're not good-tasting at all.'

The examples further confirm the necessary association of tartness (SUIMI) with sweetness, and the semantic overlap of (AMAI +) SUIMI GA ARU with AMAZUPPAI. Extralingually, these both denote the simultaneous presence of sweetness and sourness; they differ in that, with SUIMI GA ARU, the sourness is explicitly predicated as a tinge.

These findings also have implications for the meaning of SAMMI GA ARU, for whereas SUIMI GA ARU denotes a sour tinge within a sweet taste, SAMMI GA ARU by contrast denotes a sour tinge within a non-sweet taste: as we shall see in the next sub-section, these terms carry complementary compatibilities for the informant. Response patterns featuring both terms are illustrated by the following:

(B52)	Momo wa *sammi ga aru*?	Uun, *sammi* janai. *Suimi ga aru.* NO/SOURNESS/copula, negative, non-past/ TARTNESS/subject/ BE PRESENT, non-past
	'Do peaches have a sourness?'	'No, it isn't sourness. They have a tartness.'

(B53) Koogyoku wa *suimi ga aru*? — *Suimi wa nai. Sammi ga aru.*
'Do Jonathans have a tartness?' — 'They don't have a *tartness*. They have a sourness.'

where (B52) contains the copula and (B53) the particle WA, in contrast to the simple negation found with other -MI GA ARU terms as in (B54):

(B54) Sakkarin wa *sammi ga aru*? — Uun, *sammi ga nai. Nigami ga aru.*
'Does saccharine have a sourness?' — 'No, it doesn't. It has a bitterness.'

Once again, we propose that the atypical patterns in (B52) and (B53) reflect the close extralingual semantic relation between the two terms, serving rather to correct the choice of expression than to deny the relevance of the suggested parameter (i.e. sourness) outright.

5.4.2.4 Combinations of terms

We now turn to a more wide-ranging exploration of the details of combining AJI I terms. We have already seen that pairs of terms may combine, in the sense of being predicated simultaneously of a single noun. As part of our semantic investigation, we are interested in particular in the methods of combination employed in such cases and in the presence of any restrictions on the combination of individual terms.

Two syntactic patterns of combination are found, exemplified in (A17) and (A18), repeated here:

(A17) Momo wa donna aji ga suru? — *Amakute suimi ga aru.*
'What kind of a taste do peaches have?' — 'They're sweet and have a tartness.'

(A18) Sakkarin wa donna aji ga suru? — *Amai* kedo *nigami ga aru.*
'What kind of a taste does saccharine have?' — 'It's sweet but has a bitterness.'

In (A17) we find a coordinating construction involving the conjunctive (*-te*) form of the first predicator, while in (A18) we have an adversative construction with KEDO 'but'. We shall refer to these patterns as AND-combination and BUT-combination, respectively. No combinations of more than two terms have been attested in either pattern.

Possibilities of combination were investigated by testing permutations of AJI I terms for acceptability in these two frames; throughout the testing, the controlling question-frame was assumed to be (A), i.e. *X wa donna aji ga suru?* 'What kind of a taste does X have?'. The testing was conducted in two stages: first, combinations of main AJI I terms were considered. A two-way matrix was drawn up and all possible two-term combinations, in both orders, were presented to the informant and judged for acceptability as responses to question-frame (A). The matrix for this stage, with the results of testing, is set out on the following page. (In the matrix, each cell represents the combination of the two terms at whose intersection it occurs, with the terms at the left occurring first in the combination and the terms at the top second. The hatched cells occur at the intersection of each term with itself; such combinations are naturally ruled out. A blank cell indicates that the combination does not occur. Where combinations are accepted by the informant, they are written in the cells, with terms abbreviated to initial letters.)

We must recognize that a test of this kind is a somewhat blunt instrument: in particular, acceptability judgements may well be affected by considerations of real-world likelihood as well as purely linguistic factors, and we need to allow for some borderline cases among both accepted and rejected combinations. Nevertheless, the main trends are clear and can be summarized as follows:

(i) Combinations of main terms are all AND-combinations, i.e. the first term always appears in the conjunctive *-te* form.

(ii) The order of terms is irrelevant, with one exception: HORONIGAI 'pleasantly bitter' can in some cases occur in second position but not initially.

(iii) AMAI 'sweet' and the two terms which do not enter into bilateral simple negation with it, AMAZUPPAI 'sweet-sour' and AMAKARAI 'sweet-pungent', are incompatible. Moreover, as a group these three terms are strikingly restricted in combinatory power.

(iv) NIGAI 'bitter' and HORONIGAI 'pleasantly bitter', which do not enter into bilateral simple negation, are syntagmatically incompatible.

(v) In the group of terms consisting of KARAI 'pungent', its putative hyponyms SHIOKARAI 'salt' and PIRITTO(-SHITE IRU) 'piquant', and SHOPPAI 'salty' (which stands in a close extralingual

(22)

	SUPPAI	AMA-ZUPPAI	AMAI	AMA-KARAI	KARAI	SHIO-KARAI	PIRITTO (-SHITE IRU)	SHOPPAI	NIGAI	HORO-NIGAI	SHIBUI
SUPPAI 'sour'				s-te a	s-te k	s-te s	s-te p	s-te s	s-te n	s-te h	s-te s
AMAZUPPAI 'sweet-sour'										a-te chotto h	
AMAI 'sweet'										a-te chotto h	
AMAKARAI 'sweet-pungent'	a-te s						a-te p			a-te h	
KARAI 'pungent'	k-te s					k-te s		k-te s	k-te n	k-te h	k-te s
SHIOKARAI 'salt'	s-te s				s-te k		s-te p		s-te n	s-te h	s-te s
PIRITTO (-SHITE IRU) 'piquant'	p-te s			p-te a		p-te s		p-te s	p-te n	p-te k	p-te s
SHOPPAI 'salty'	s-te s				s-te k		s-te p		s-te n	s-te h	s-te s
NIGAI 'bitter'	n-te s				n-te k	n-te s	n-te p	n-te s			n-te s
HORONIGAI 'pleasantly bitter'	h-te s						h-te p	h-te s			
SHIBUI 'astringent'	s-te s				s-te k	s-te s	s-te p	s-te s	s-te n	s-te h	

relationship with SHIOKARAI), the matrix shows, as expected, that PIRITTO(-SHITE IRU) is incompatible with KARAI and that SHOPPAI is incompatible with SHIOKARAI. A major deviation from the proposed analysis so far, however, is that the combination of KARAI 'pungent' and SHIOKARAI 'salt' is permitted; since, by definition, a superordinate term and a hyponym cannot combine in this fashion, the status of these two terms must now be reconsidered.

First, for convenience let us repeat the relevant examples for KARAI 'pungent' and SHIOKARAI 'salt':

(B24)	Shooyu wa *shiokarai*? 'Is soy sauce salt?'	Uun, *shiokaraku* nai. Shooyu wa *karai* dake. 'No, it isn't. Soy sauce is just pungent.'
(B25)	Umi no mizu wa *karai*? 'Is sea-water pungent?'	Un, *karai*. *Shiokarai*. 'Yes, it is. It's salt.'

As noted earlier, these examples clearly suggest a relationship of hyponymy between KARAI, as superordinate, and SHIOKARAI. The present evidence, however, shows that such an analysis cannot be correct as it stands. A hypothesis that would account for both aspects of the data is that KARAI in fact functions at two levels of generality, subsuming SHIOKARAI in one case but not in the other: i.e. that there is not one sense of KARAI but two senses, which we may refer to as $KARAI_1$ and $KARAI_2$, in a structure that can be represented as follows:

(23)

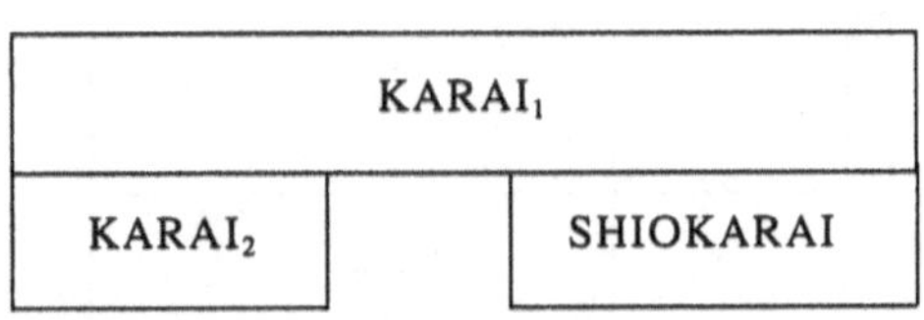

with SHIOKARAI a hyponym of $KARAI_1$ but at the same contrast level with $KARAI_2$.

In seeking evidence to support this analysis, we may first return to consider extralingual meaning, where it is noticeable that the application of KARAI covers a range of readily discriminable taste qualities. As we have seen, focal exemplars include hot substances such as mustard (*karashi*), red peppers (*toogarashi*), curry powder (*kareeko*), pepper (*koshoo*), *wasabi* horse-radish (*wasabi*) and ginger (*shooga*), together with such substances as soy sauce (*shooyu*) and salt (*shio*). It is not difficult to imagine a substance which combines different taste qualities from within this range: according to the informant, for example, a salty type of curry would qualify for description as *karakute shiokarai.* This combination must involve $KARAI_2$, which points to curry (*karee*), a hot substance, as falling within the range of application of this term. The wide denotational range of KARAI is also reflected in the fact that, after a response in which it occurs, such as in the following example:

(A35)	Hagisu wa donna aji ga suru?	*Karai.*
	'What kind of a taste does haggis have?'	'It's pungent.'

one can go on to ask the natural question:

(24) Donna *karasa*?
'What kind of pungency is it?'

and appropriate responses include *Shiokarai* 'It's salt', *Kareeko no karasa* 'It's the pungency of curry powder', *Shooyu no karasa* 'It's the pungency of soy sauce', *Piritto-shiteru* 'It's piquant', etc. (Cf. also the form of the response to (B34) in 5.4.2.2: *(Shooyu wa) piritto-shita karasa janai* '(Soy sauce) isn't a piquant pungency'.)

Outside of the primary data, it is relevant to note that definitions of the taste term KARAI in Japanese dictionaries are generally given in two separate sub-entries (occasionally, in two entries), one covering the taste of hot substances, the other covering a salt taste. Thus Hayashi et al. (1987: 189) define these senses of KARAI as (1) *Toogarashi ya kareeko o kuchi ni ireta toki no yoo na, shita ga hirihiri suru yoo na kanji da* ('Producing a burning sensation of the tongue, as when one puts red peppers or curry powder in one's mouth'), (2) *Shioke ga tsuyoi* ('Having a strong saltiness').[18] Also relevant here are forms such as TOOGARASHIKARAI, NAMBANKARAI and KOSHOOKARAI reported for certain dialects alongside KARAI in the Linguistic Atlas of Japan survey (Kokuritsu Kokugo Kenkyuujo, 1966a: 25): the presence of these specific compound terms denoting hot tastes (in the dialects concerned, TOOGARASHI,

NAMBAN and KOSHOO all denote red peppers) clearly reflects a perceived need for finer semantic discriminations in this area.

Intralingually, we find that questions relating to different food-substances give rise to different response patterns involving SHIOKARAI and KARAI:

(B24)	Shooyu wa *shiokarai*? 'Is soy sauce salt?'	Uun, *shiokaraku* nai. Shooyu wa *karai* dake. 'No, it isn't. Soy sauce is just pungent.'
(B55)	Karee wa *shiokarai*? 'Is curry salt?'	Uun, karee wa *karai*. 'No, curry is pungent.'
(B56)	Shooga wa *shiokarai*? 'Is ginger salt?'	Shooga wa *shiokaraku* nai. Shooga wa *karai*. 'No, it isn't. Ginger is pungent.'
(B57)	Karashi wa *shiokarai*? 'Is mustard salt?'	Uun, karashi wa *karai*. 'No, mustard is pungent.'

In response to the same question-frame (*X wa shiokarai?* 'Is X salt?'), (B24), with SHOOYU 'soy sauce,' contains DAKE 'only, just' in the reply, while (B55)–(B57), with KAREE 'curry', SHOOGA 'ginger' and KARASHI 'mustard', do not. These patterns are consistent, and attempts to interchange them are rejected by the informant. These results are in keeping with the view that we are dealing with two different senses of KARAI: we conclude that (B24), with DAKE 'only, just', involves KARAI$_1$, the superordinate of SHIOKARAI 'salt', while (B55)–(B57), without DAKE, involve KARAI$_2$. Moreover, the extralingual distinction referred to above is maintained in these examples: curry, ginger and mustard are all hot in taste.

We conclude that the restructuring proposed above can be supported on both intralingual and extralingual grounds. From this point on we maintain the gloss 'pungent' for the more general KARAI$_1$; for KARAI$_2$, 'hot' is an appropriate equivalent.

This new analysis necessitates a re-examination of the status of PIRITTO (-SHITE IRU) 'piquant'. To repeat the examples considered thus far:

(B33)	Shooyu wa *piritto-shiteru*? 'Is soy sauce piquant?'	Uun, *piritto-shitenai.* Shooyu wa *karai* dake. 'No, it isn't. Soy sauce is just pungent.'

and

(B30)	Karashina wa *karai*? 'Is mustard-plant pungent?'	Un, chotto *piritto-shiteru* ne. 'Yes, it's somewhat piquant.'

were interpreted as indicating a relationship of hyponymy between KARAI (as superordinate) and PIRITTO(-SHITE IRU). Given the distinction between KARAI_1 'pungent' and KARAI_2 'hot', however, the question now arises of how PIRITTO(-SHITE IRU) relates to each of these senses. (B33) can reasonably be interpreted as containing KARAI_1 'pungent', on the basis of the presence of the item SHOOYU 'soy sauce'; PIRITTO(-SHITE IRU) may then be seen as a hyponym of that term, on the strength of the occurrence of DAKE. It remains, therefore, to examine the relationship of PIRITTO(-SHITE IRU) to KARAI_2 'hot'. If we construct questions containing items denoting hot food-substances, as follows:

(B58)	Karee wa *piritto-shiteru*? 'Is curry piquant?'	*Piritto-shiteru* dokoroka *karai.* 'Far from being just piquant, it's hot.'
(B59)	Karashi wa *piritto-shiteru*? 'Is mustard piquant?'	Soo-ne, chotto taberu to ne. Ammari taberu to *karai.* 'Well, it is if you just eat a little. If you eat a lot, it's hot.'

we find different response patterns from those above. Replies with DAKE 'only, just' to (B58) and (B59) are rejected by the informant, and this appears to confirm that we are here dealing with KARAI_2. The construction with DOKOROKA in (B58) has already been seen with SHOPPAI 'salty' and SHIOKARAI 'salt' (cf. (B26), 5.4.2.2): it expresses a strong denial of the preceding predication, often in favour of a more extreme alternative. As with SHIOKARAI and SHOPPAI, it may be reasonably interpreted as reflecting a close extralingual relation, the two terms denoting differing intensities of the same quality; i.e. PIRITTO (-SHITE IRU) 'piquant' denotes a lesser, and KARAI$_2$ 'hot' a more extreme, degree of hotness, and this analysis is supported by the response to (B59). Unlike in the present case, however, with SHIOKARAI and SHOPPAI we found simple negation in the reverse direction (cf. (B27)), and we thus analysed the pair intralingually as contrast terms. Here we find assertion in the reverse direction, as seen in (B30), and this clearly indicates a relation of hyponymy.

The relevant section of the structural diagram at the end of 5.4.2.2 may now be amended as follows to take account of these relationships:

(25)

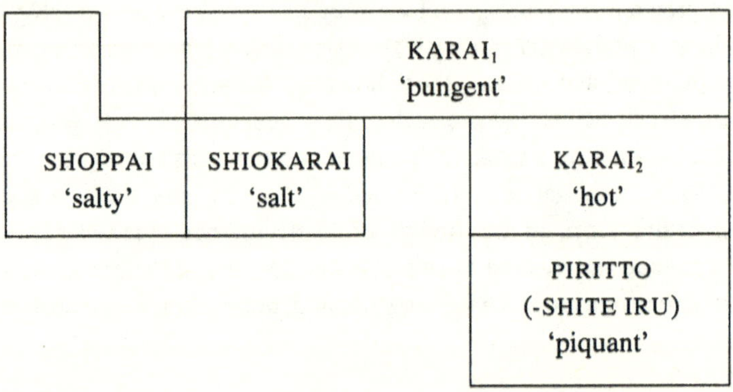

Thus, KARAI$_1$ 'pungent' is the superordinate term of which KARAI$_2$ 'hot' and SHIOKARAI 'salt' are immediate hyponyms. At the same time PIRITTO(-SHITE IRU) is a hyponym of KARAI$_2$ and thereby (since hyponymy is a transitive relation) also of KARAI$_1$. It is important to note that the denotation of its hyponyms does not exhaust the total denotational range of KARAI$_1$: the tastes of certain substances, notably soy sauce (*shooyu*) but also bean paste (*miso*), are described as KARAI$_1$ without falling within the range of either KARAI$_2$ or SHIOKARAI, the type of pungency involved in such cases being capable of specification, where necessary, by syntactic means (*shooyu no karasa* 'the pungency of soy sauce', etc.). This state of affairs is not unusual: with English colour terms, for example, the denotational range of terms such as SCARLET and CRIMSON does not exhaust that of the superordinate, RED, and certain shades remain to be characterized by expressions such as *light red*, *just red*, etc.

The combinatorial matrix for main AJI I terms presented earlier must also be enlarged to include both senses of KARAI: KARAI$_1$ and KARAI$_2$ will have identical entries except for the fact that KARAI$_1$ does not combine with its hyponym SHIOKARAI. The relevant sections of the matrix will thus be as follows:

(26)

	KARAI$_1$	KARAI$_2$	SHIO-KARAI	PIRITTO (-SHITE IRU)	SHOPPAI
KARAI$_1$ 'pungent'					k-te s
KARAI$_2$ 'hot			k-te s		k-te s
SHIO-KARAI 'salt'		s-te k		s-te p	
PIRITTO (-SHITE IRU) 'piquant'			p-te s		p-te s
SHOPPAI 'salty'	s-te k	s-te k		s-te p	

The second stage of testing involved combinations of main terms and -MI GA ARU terms. Once again, the results may be summarized in matrix form:

(27)

	SUPPAI 'sour'	AMA-ZUPPAI 'sweet' -sour'	AMAI 'sweet'	AMA-KARAI 'sweet-pungent'	$KARAI_1$ 'pungent'	$KARAI_2$ 'hot'	SHIO-KARAI 'salt'	PIRITTO (-SHITE IRU) 'piquant'	SHOPPAI 'salty'	NIGAI 'bitter'	HORO-NIGAI 'pleas-antly bitter'	SHIBUI 'astrin-gent'
SAMMI GA ARU 'has a sourness'				a-te s ga aru	k-te s ga aru	k-te s ga aru	s-te s ga aru	p-te s ga aru	s-te s ga aru	n-te s ga aru	h-te s ga aru	s-te s ga aru
SUIMI GA ARU 'has a tartness'			a-te s ga aru									
AMAMI GA ARU 'has a sweetness'	s kedo a ga aru				k kedo a ga aru	k kedo a ga aru	s kedo a ga aru	p kedo a ga aru	s kedo a ga aru	n kedo a ga aru	h-te a ga aru	s kedo a ga aru
NIGAMI GA ARU 'has a bitterness'	s-te n ga aru	a kedo n ga aru	a kedo n ga aru	a kedo n ga aru	k-te n ga aru	k-te n ga aru	s-te n ga aru	p-te n ga aru	s-te n ga aru			s-te n ga aru
SHIBUMI GA ARU 'has an astringency'	s-te s ga aru	a kedo s ga aru	a kedo s ga aru	a kedo s ga aru	k-te s ga aru	k-te s ga aru	s-te s ga aru	p-te s ga aru	s-te s ga aru	n-te s ga aru	h-te s ga aru	

(The conventions are similar to those in the earlier matrix: Each cell represents the combination of the ordinary term and the -MI GA ARU term at whose intersection it occurs. In all cases -MI GA ARU terms occur second in the combination.)

The main patterns illustrated in this matrix can be summarized as follows:

(i) Both AND- and BUT-combinations occur; i.e. the first term appears in some cases in the conjunctive *-te* form, in other cases followed by KEDO 'but'.
(ii) The natural order is main term + -MI GA ARU term.
(iii) SAMMI GA ARU 'has a sourness' and SUIMI GA ARU 'has a tartness' are complementary in their combinatory power: apart from AMAZUPPAI 'sweet-sour', which is incompatible with both terms, SAMMI GA ARU combines with all terms except AMAI 'sweet', while SUIMI GA ARU combines only with AMAI (cf. 5.4.2.3).
(iv) NIGAMI GA ARU 'has a bitterness' and SHIBUMI GA ARU 'has an astringency' are unrestricted in combinatory power except that NIGAMI GA ARU and HORONIGAI 'pleasantly bitter' are incompatible.
(v) AMAI 'sweet' and the two terms which do not enter into bilateral simple negation with it, AMAZUPPAI 'sweet-sour' and AMAKARAI 'sweet-pungent', feature in BUT-combination with NIGAMI GA ARU 'has a bitterness' and SHIBUMI GA ARU 'has an astringency'.
(vi) AMAMI GA ARU 'has a sweetness' is incompatible with AMAZUPPAI and AMAKARAI. With the exception of HORONIGAI 'pleasantly bitter', all combinations into which it enters are BUT-combinations.

Once again, the matrix reveals interesting differences in combinatory freedom. Moreover, among combinations that are permitted, some terms are opposed (i.e. appear in BUT-combination) in a way that others are not.

On the basis of the two matrices, we may set up groupings of mutually incompatible terms. To begin with clear cases, the following initial groupings follow naturally from our semantic findings thus far:

Group S: SUPPAI 'sour'/SAMMI GA ARU 'has a sourness'
Group A: AMAI 'sweet'/AMAMI GA ARU 'has a sweetness'
Group N: NIGAI 'bitter'/NIGAMI GA ARU 'has a bitterness'
Group SH: SHIBUI 'astringent'/SHIBUMI GA ARU 'has an astringency'

We have concluded in 5.4.2.3 that main terms and their associated -MI GA ARU terms predicate essentially the same quality as a predominant taste and as a tinge taste, respectively; since a given taste quality cannot be present in both these capacities simultaneously, these incompatibilities follow as a matter of course. A further group, group K, contains linked sets of mutually incompatible terms:

Group K: KARAI$_1$ 'pungent'/KARAI$_2$ 'hot'/PIRITTO(-SHITE IRU) 'piquant'
KARAI$_1$ 'pungent'/SHIOKARAI 'salt'

The incompatibilities here rest on the intralingual relations of hyponymy holding within the sets concerned.

To these we may add the following cases where incompatibility is reasonably explained in terms of overlap in extralingual meaning:

SUIMI GA ARU 'has a tartness'/SUPPAI 'sour'

AMAZUPPAI/ 'sweet-sour' <
AMAI 'sweet'/AMAMI GA ARU 'has a sweetness'
SUPPAI 'sour'/SAMMI GA ARU 'has a sourness'/ SUIMI GA ARU 'has a tartness'

AMAKARAI/ 'sweet-pungent' <
AMAI 'sweet'/AMAMI GA ARU 'has a sweetness'
KARAI$_1$ 'pungent'

HORONIGAI 'pleasantly bitter'/NIGAI 'bitter'/NIGAMI GA ARU 'has a bitterness'

SHOPPAI 'salty'/SHIOKARAI 'salt'

These incompatibilities are predictable on the basis of shared denotational parameters of sourness (SUIMI GA ARU: cf. 5.4.2.3), sweetness and sourness (AMAZUPPAI), sweetness and pungency (as of soy sauce) (AMAKARAI), bitterness (HORONIGAI), and saltiness (SHOPPAI). Intralingually, there is no evidence for hyponymy among these terms; however, with precisely these items, there have been other signs of semantic affinity. Thus, for SHOPPAI and SHIOKARAI we have seen the occurrence of DOKOROKA in one direction, and elsewhere we have noted in some responses the presence of the particle WA (together with KEDO 'but') serving to attenuate negation (cf. 5.4.2.1) or to add a concessive tone to an assertion (cf. 5.4.2.3):

(B42)	Momo wa *suppai*?	*Suppaku nai* kedo *suimi wa aru.*
	'Are peaches sour?'	'No, they're not, but they do have a tartness.'
(B11)	Mikan wa *amai*?	*Amaku* wa nai kedo *amazuppai.*
	'Are mandarins sweet?'	'They're not *sweet*, but they're sweet-sour.'
(B21)	Sukiyaki wa *amai*?	Soo-ne, *amaku* wa nai kedo *amakarai.*
	'Is *sukiyaki* sweet?'	'Well, it's not *sweet*, but it's sweet-pungent.'
(B23)	Chokoreeto no burakku wa *nigai*?	*Nigaku* wa nai kedo *horonigai.*
	'Is plain chocolate bitter?'	'It's not *bitter*, but it's pleasantly bitter.'

These responses provide some intralingual support for a particular closeness of meaning between SUIMI GA ARU and SUPPAI, AMAI and AMAZUPPAI, AMAI and AMAKARAI, and NIGAI and HORONIGAI. Interestingly, for the compound terms AMAZUPPAI and AMAKARAI, responses with attenuated negation are found only with AMAI; with SUPPAI and KARAI respectively, despite the equally clear denotational overlap, we find bilateral simple negation. This is clearly a point of some semantic significance, which is not captured by a description of these terms simply as 'mixtures': in some sense they are semantically closer to AMAI, and on this basis we shall assign them to Group A. As we shall see below, this grouping is borne out in terms of compatibilities observed across groups. The expanded groups, consisting of terms which *prima facie* fall together on intralingual or extralingual semantic grounds, are as follows:

Group S: SUPPAI/SAMMI GA ARU/SUIMI GA ARU
Group A: AMAI/AMAZUPPAI/AMAKARAI/AMAMI GA ARU
Group N: NIGAI/HORONIGAI/NIGAMI GA ARU
Group SH: SHIBUI/SHIBUMI GA ARU
Group K: $KARAI_1$/$KARAI_2$/PIRITTO(-SHITE IRU)
$KARAI_1$/SHIOKARAI
SHIOKARAI/SHOPPAI

Within these groups, then, incompatibility is to be expected on the basis of what we have learned about the meaning of the terms involved. Superimposed upon these natural incompatibilities, however, we find some cases of incompatibility across groups (cf. 5.4.2.1). Since cross-group combination might *a priori* be expected to be free, such restrictions obviously reflect an important aspect of the semantic structure of AJI I.

The general nature of the cross-group restrictions for main terms is clearly brought out in the following diagram. The letters stand for the groups of terms above; solid lines between groups represent free combination, broken lines restricted combination.

(28)

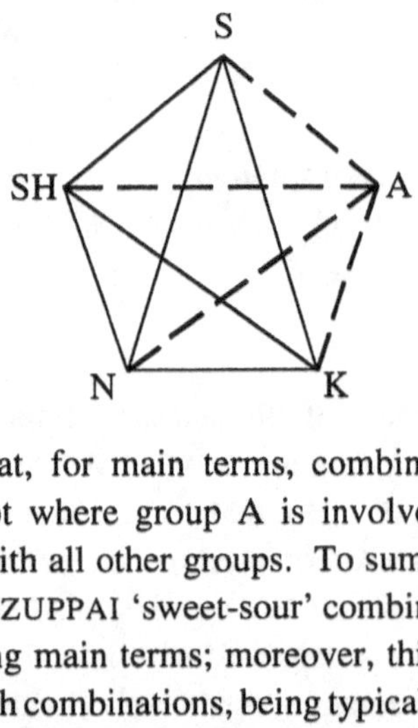

The diagram shows that, for main terms, combination between groups is everywhere free, except where group A is involved: for group A, indeed, restrictions are found with all other groups. To summarize these restrictions, AMAI 'sweet' and AMAZUPPAI 'sweet-sour' combine only with HORONIGAI 'pleasantly bitter' among main terms; moreover, this term apparently carries subordinate status in such combinations, being typically modified by the adverb CHOTTO 'somewhat' and standing in second position. AMAKARAI 'sweet-pungent' is somewhat freer, combining with HORONIGAI (as a second member) and with SUPPAI 'sour' and PIRITTO(-SHITE IRU) 'piquant'; nevertheless it does not combine with NIGAI 'bitter' or SHIBUI 'astringent'.

Turning to combinations of main and -MI GA ARU terms, combination across groups here is essentially free; the only complications involve the complementary behaviour of SAMMI GA ARU 'has a sourness' and SUIMI GA ARU 'has a tartness' previously noted, and the predictable incompatibility with group S of AMAZUPPAI 'sweet-sour'. The contrast with main + main term combinations for group A is thus striking: here, AMAMI GA ARU 'has a sweetness' is freely compatible with all main terms, and AMAI 'sweet' itself combines with -MI GA ARU terms from all groups. At the same time, however, -MI GA ARU terms introduce the new distinction between AND-combination and BUT-combination, which in turn reflects a semantic division amongst individual terms. Once again, it is useful to illustrate this diagrammatically, using solid lines here to represent AND-combination and broken lines the additional presence of BUT-combination:

(29)

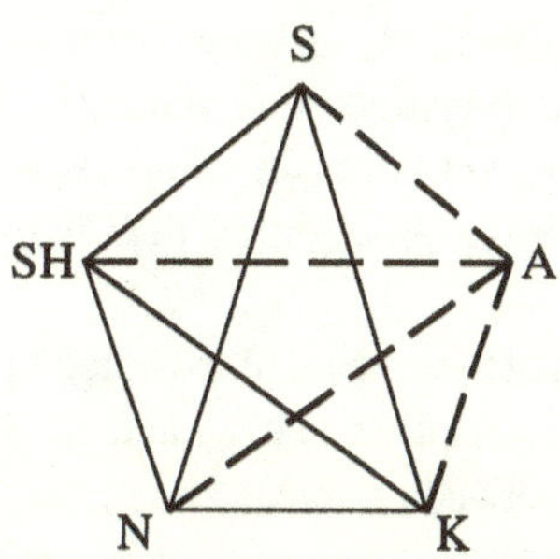

As before, group A stands out clearly: AND-combination holds throughout, except with group A terms, where we find both combination types; moreover, BUT-combinations are present between group A and all other groups. Thus the following all involve BUT-combination:

AMAI 'sweet'/AMAZUPPAI 'sweet-sour'/AMAKARAI 'sweet-pungent' + NIGAMI GA ARU 'has a bitterness'/SHIBUMI GA ARU 'has an astringency'

AMAMI GA ARU 'has a sweetness' + SUPPAI 'sour'/KARAI_1 'pungent'/ KARAI_2 'hot'/PIRITTO(-SHITE IRU) 'piquant'/SHIOKARAI 'salt'/ SHOPPAI 'salty'/NIGAI 'bitter'/SHIBUI 'astringent'

The only AND-combinations involving group A terms are the following:

AMAI 'sweet' + SUIMI GA ARU 'has a tartness'
AMAKARAI 'sweet-pungent' + SAMMI GA ARU 'has a sourness'
AMAMI GA ARU 'has a sweetness' + HORONIGAI 'pleasantly bitter'

We conclude that, uniquely, group A terms stand semantically opposed to the majority of AJI I terms, the sole exceptions being HORONIGAI 'pleasantly bitter' among main terms and -MI GA ARU terms from group S (SUIMI/SAMMI GA ARU 'has a tartness/sourness').

Summarizing the evidence from combinatorial behaviour, it is clear that group A terms stand out as occupying a special place within AJI I. In the first place, in main term combinations their compatibility is highly restricted. In this connection it is noteworthy that the two compounds formed by combining taste term roots, namely AMAZUPPAI 'sweet-sour' and AMAKARAI 'sweet-pungent', both involve AMAI, and these items may plausibly be seen as functioning to lexicalize complex taste qualities which cannot be described by syntactic combinations of AMAI + SUPPAI or AMAI + $KARAI_1$ respectively. Secondly, in main + -MI GA ARU combinations they are the only terms which exhibit BUT-combination and thereby demonstrate their semantic opposition to the majority of AJI I terms; furthermore, AMAI has the unique characteristic of combining with SUIMI GA ARU 'has a tartness', rather than with SAMMI GA ARU 'has a sourness', from group S.

This serves to remind us of the special place noted in 1.2 for 'sweet' in Korean on the basis of Maeda's description (1978) and in Galelan by Ishige (1983). Apropos of the opposition between group A and (most) other AJI I terms, we may reflect that, in English too, SWEET seems to stand opposed to other taste terms in that such pairs as SWEET/SOUR and SWEET/BITTER are felt as opposites more strongly than, say, SOUR/BITTER. In Japanese, the major opposition in this sense is undoubtedly felt to be AMAI/KARAI and, as we shall see below (6.4), these terms indeed function as true antonyms in certain extended fields of meaning.

5.4.2.5 Affective value

It is natural to enquire whether the facts of combination noted in the previous sub-section can be related to any other demonstrable semantic property of AJI I terms. One that suggests itself is the affective value associated with many taste terms (cf. 1.1, 1.3).

A linguistic basis for an affective dimension of meaning may be established in terms of combination with the evaluative taste terms OISHII/UMAI 'good-tasting' and MAZUI 'bad-tasting' discussed in Chapter 4. We have already seen some combinations of this type in response to the question-frames *X wa oishii/umai?* 'Is/are X good-tasting?' (4.1) and *Sono X wa doo?* 'How is that X?' (4.6). At the generic level which is our concern here, the former is the more appropriate frame and, as illustrated below, we find that different AJI I terms indeed show different patterns of combination. The following are examples of typical question-and-answer pairs:

	Question	Answer
(30)	Momo wa oishii? PEACH/isolator/ GOOD-TASTING, non-past 'Are peaches good?'	Un, oishii – *amakute oishii.* YES/GOOD-TASTING, non-past/SWEET, conjunctive/GOOD-TASTING, non-past 'Yes, they are – they're sweet and (thereby) good.'
(31)	Chokoreeto no burakku wa oishii? 'Is plain chocolate good?'	Un, *horonigakute oishii.* 'Yes, it's pleasantly bitter and (thereby) good.'
(32)	Kobucha wa oishii? 'Is *kobucha* (tea) good?'	Un, chotto *shoppai* kedo *oishii.* YES/SOMEWHAT/SALTY, non-past/BUT/GOOD-TASTING, non-past 'Yes, it's a bit salty but it's good.'
(33)	Umeboshi wa oishii? 'Are pickled plums good?'	Un, umeboshi wa *suppai* kedo *oishii.* 'Yes, they're sour but they're good.'
(34)	Madorasukaree wa oishii? 'Is Madras curry good?'	Soo-ne, *karai* kedo *oishii.* 'Well, it's hot but, yes, it's good.'
(35)	Umi no mizu wa oishii? 'Is sea-water good?'	Uun, umi no mizu wa *karakute mazui.* 'No, it's pungent and (thereby) tastes bad.'

(36)	Shibugaki wa oishii?	Shibugaki wa *mazui* yo – *shibui* kara.[19]
		SHIBUGAKI/isolator/ BAD-TASTING, non-past/ emphasizer/ASTRINGENT, non-past/BECAUSE
	'Are *shibugaki* good?'	No, they taste bad – because they're astringent.'

The patterns of combination found in such responses can be divided into three types, as follows:

(a) *X-te oishii/umai* 'is X and is (thereby) good-tasting'[20] (cf. (30), (31) above)
(b) *X kedo oishii/umai* 'is X but is good-tasting' (cf. (32)–(34))
(c) *X-te mazui/X kara mazui* 'is X and (thereby) bad-tasting'/'is bad-tasting because is X' (Cf. (35), (36))

AJI I terms fall into three groups, occurring in pattern (a) alone, in patterns (b) and (c), or in pattern (c) alone; for convenience, we refer to these groupings as +OISHII, (–)OISHII, and –OISHII, respectively:

(37)

+OISHII	(–)OISHII	–OISHII
AMAI 'sweet'	SUPPAI 'sour'	SHIOKARAI 'salt'
AMAZUPPAI 'sweet-sour'	KARAI$_1$ 'pungent'	NIGAI 'bitter'
AMAKARAI 'sweet-pungent'	KARAI$_2$ 'hot'	SHIBUI 'astringent'[21]
AMAMI GA ARU 'has a sweetness'	SHOPPAI 'salty'	
SUIMI GA ARU 'has a tartness'	NIGAMI GA ARU 'has a bitterness'	
SAMMI GA ARU 'has a sourness'	SHIBUMI GA ARU 'has an astringency'	
PIRITTO (-SHITE IRU) 'piquant'		
HORONIGAI 'pleasantly bitter'		

Characterized as +OISHII are thus all group A terms, together with SUIMI GA ARU, SAMMI GA ARU, PIRITTO(-SHITE IRU) and HORONIGAI, and it is clear that there is a close, if not perfect, correlation between the affective value of AJI I terms and their behaviour in combination. Let us recall that in main term combinations HORONIGAI is uniquely compatible with all group A terms, and that in main + tinge term combinations we find only HORONIGAI, SUIMI GA ARU and SAMMI GA ARU in AND-combination with group A. In addition, PIRITTO(-SHITE IRU) combines with AMAKARAI which, however, also appears to combine with SUPPAI, a (–)OISHII term. With this last exception, we may say that group A terms demand a shared +OISHII characterization in these types of combination; this is a necessary, but not sufficient, condition in that PIRITTO (-SHITE IRU) from group K is generally excluded, except with AMAKARAI.

Looking at the findings from the standpoint of the overall system, group A terms are unique in having the same characterization (+OISHII) throughout, with AMAZUPPAI and AMAKARAI again falling together in this respect with AMAI rather than with SUPPAI or KARAI. The other groups all contain mixed terms: group S straddles +/(–)OISHII, groups K and N straddle all three categories, and group SH straddles (–)/–OISHII. In pairs of terms whose denotation differs in intensity, we note that the lesser members (PIRITTO(-SHITE IRU), SHOPPAI) occupy a higher position (i.e. closer to +OISHII) than their extreme partners ($KARAI_2$, SHIOKARAI). We also note that -MI GA ARU terms generally occupy a higher position than their main term counterparts, except of course in group A where all terms are +OISHII. The results conform well with our everyday intuitions concerning the affective value of taste qualities. Sweet and bitter, for instance, are the qualities most commonly cited in this connection, and AMAI and NIGAI are clearly characterized as +OISHII and –OISHII respectively; the disposition of degree and tinge terms also supports the observation that non-sweet substances tend to be perceived as unpleasant with rising concentrations (cf. 1.1). Apropos of this, let us also note that the four +OISHII terms outside group A all denote attenuated taste qualities: PIRITTO(-SHITE IRU) denotes a limited degree of hotness (vis-a-vis $KARAI_2$), HORONIGAI a pleasant type of bitterness, and SUIMI GA ARU and SAMMI GA ARU tinges of sourness.

We conclude that there are linguistic grounds for postulating an affective value as part of the meaning of AJI I terms (cf. 1.3). Thus, +OISHII terms denote taste qualities which, at a general level, are perceived as pleasant, other terms qualities which are perceived as basically unpleasant, strongly so in the case of –OISHII terms. Let us stress once again, however, that our focus here is on the general, or normal, level. As we have seen, the fact that AMAI 'sweet' denotes a basically pleasant taste quality does not prevent sweetness from being an undesirable attribute in specific cases where taste norms are infringed. Neither does it prevent individuals from developing personal tastes which do not favour sweet substances, or from finding them undesirable on particular occasions.

5.4.3 Semantics of individual terms

Below we bring together our main conclusions concerning the terms of AJI I.

Beginning with main terms, we conclude that all terms are contrast terms with the exception of the pairs KARAI$_1$/SHIOKARAI and (KARAI$_1$/)KARAI$_2$/PIRITTO (-SHITE IRU), where we find relations of hyponymy. As argued in 5.4.2.1, our criterion for the relation of contrast is based on the permissibility of simultaneous negation and assertion in patterns such as *(A) is not X, it's Y/(B) is not Y, it's X.* The simplest cases in the data involve bilateral simple negation, which holds between the majority of AJI I main term pairs. For SHIOKARAI/SHOPPAI we find negation combined with the use of DOKOROKA ('far from'), indicating, in addition to contrast, a close extralingual relationship, differing only in intensity. (With KARAI$_2$/PIRITTO(-SHITE IRU) we likewise find DOKOROKA used in one direction and assume a similar extralingual relationship in addition to hyponymy.) With the remaining pairs, AMAI/AMAZUPPAI, AMAI/AMAKARAI and NIGAI/HORONIGAI, we have found simple negation in one direction and WA-negation in the other and have noted this as an indicator of semantic closeness. Intralingually, we conclude that these are simple contrast pairs; however, close extralingual links are apparent in that NIGAI/HORONIGAI both denote the presence of bitterness, and AMAI/AMAZUPPAI/AMAKARAI all denote the presence of sweetness and, additionally, share positive affective value. These relations for main terms may be diagrammed as follows:

(38)

<table>
<tr><td rowspan="3">SUPPAI</td><td rowspan="3">AMA-
ZUPPAI</td><td rowspan="3">AMAI</td><td rowspan="3">AMA-
KARAI</td><td rowspan="3">SHOPPAI</td><td colspan="2">KARAI$_1$</td><td rowspan="3">NIGAI</td><td rowspan="3">HORO-
NIGAI</td><td rowspan="3">SHIBUI</td></tr>
<tr><td rowspan="2">SHIO-
KARAI</td><td>KARAI$_2$</td></tr>
<tr><td>PIRITTO
(-SHITE
IRU)</td></tr>
</table>

Here, terms separated by vertical lines are contrast terms, and terms separated by a horizontal line are related by hyponymy, with superordinate terms above their hyponyms.

Next we consider -MI GA ARU terms (cf. 5.4.2.3). As a group, these constitute a set of contrast terms. Individually, all terms except SUIMI GA ARU relate clearly to particular main terms, with which they basically exhibit patterns of bilateral assertion; we may represent this by placing -MI GA ARU terms together with their corresponding main terms in the same box. With regard to SUIMI GA ARU, we have seen that this term shows negative patterns with SUPPAI, albeit with reflections of denotational similarity, and we have concluded that this results from the special place of SUIMI GA ARU vis-a-vis AMAI: whereas -MI GA ARU terms in general predicate taste qualities as tinges, SUIMI GA ARU additionally carries the definite implication that the predominant taste quality is sweet. We therefore list it under SUPPAI with SAMMI GA ARU; both these -MI GA ARU terms denote the presence of a sour tinge, but in complementary situations, SUIMI GA ARU where there is a predominant sweet taste, SAMMI GA ARU elsewhere:

(39)

<table>
<tr><td colspan="2" rowspan="2">SUPPAI</td><td rowspan="3">AMA-ZUPPAI</td><td rowspan="2">AMAI</td><td rowspan="3">AMA-KARAI</td><td rowspan="3">SHOP-PAI</td><td colspan="2">KARAI$_1$</td><td rowspan="2">NIGAI</td><td rowspan="3">HORO-NIGAI</td><td rowspan="2">SHIBUI</td></tr>
<tr><td rowspan="2">SHIO-KARAI</td><td>KARAI$_2$</td></tr>
<tr><td>SAMMI GA ARU</td><td>SUIMI GA ARU</td><td>AMAMI GA ARU</td><td>PIRITTO (-SHITE IRU)</td><td>NIGAMI GA ARU</td><td>SHIBUMI GA ARU</td></tr>
</table>

In 5.4.2.4 we found that syntagmatic compatibility varies among terms, and we may build the representation of these relations into the diagram in two stages. Defining incompatibility as inability to co-occur in AND-combination, we may first modify the diagram as follows:

(40)

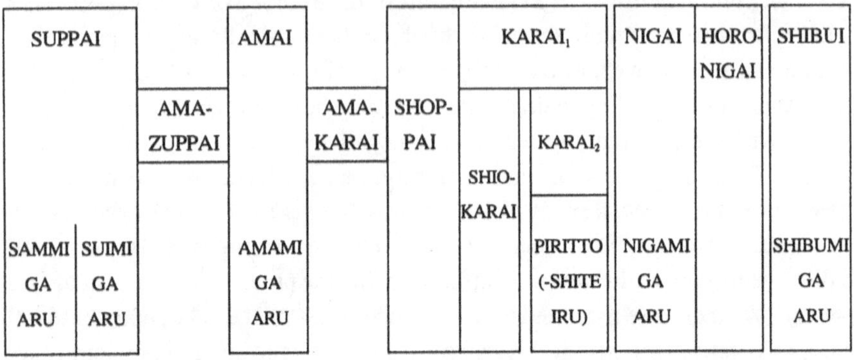

Here adjoining terms are incompatible, while non-adjoining terms are in principle compatible. The diagram in effect reproduces the groups isolated in 5.4.2.4 with, additionally, the role of AMAZUPPAI and AMAKARAI as bridging terms between groups (A and S, and A and K, respectively) being clearly indicated. Finally, superimposed on this structure there is a further set of incompatibilities centred on AMAI and other group A terms. These may be represented as follows:

(41)

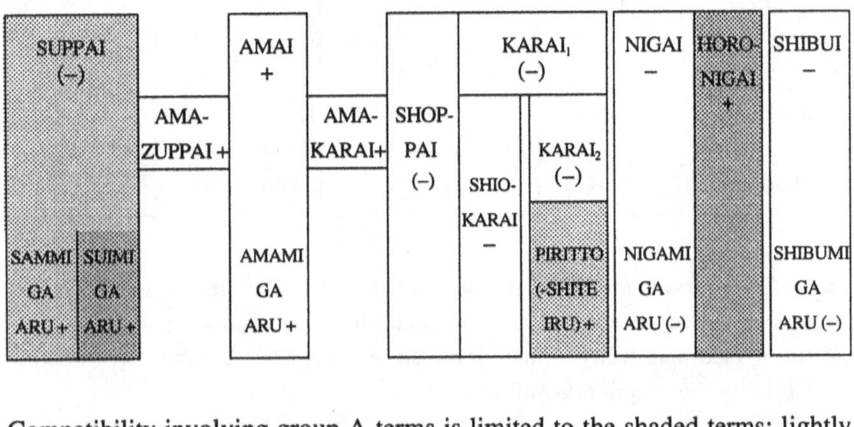

Compatibility involving group A terms is limited to the shaded terms; lightly shaded terms are compatible only with AMAKARAI, which has slightly wider combinative possibilities. Affective values based on combinative behaviour with evaluative taste terms (cf. 5.4.2.5) have been added: group A terms themselves are all characterized as +OISHII, and their strong preference for combination with like-valued terms is clearly evident.

6 *Descriptive taste terms in Japanese (II)*

In this chapter we continue our treatment of the semantics of the lexical field of AJI. The remaining systems, AJI II and AJI III, are analysed in 6.1 and 6.2, and the overall field is reviewed in 6.3. Finally, extended meanings of descriptive taste terms in Japanese are surveyed in 6.4.

6.1 The lexical system AJI II

AJI II contains two terms, AKUPPOI 'harsh' and AKU GA ARU 'have a harshness'.

6.1.1 Extralingual meaning

Focal exemplars given by the informant for both terms are GOBOO 'burdock', NASU 'aubergine' and SATOIMO 'taro'. The dictionaries surveyed in 5.4.1 all define AKU as a component of plants/vegetables, though without reference to specific examples. One dictionary (Kindaichi et al., 1989: 11) defines AKU as *shokubutsu ni fukumareru, egui seibun* 'the harsh (EGUI) component found in plants'; as we have seen in 5.2, EGUI and associated terms are part of the taste vocabularies of some speakers but are not used by the informant, and this area is clearly characterized by considerable variation (cf. Chapter 5, note 8). Other dictionaries refer to astringency (SHIBUMI) in plants: although the present terms and SHIBUI belong to separate systems for the informant, this suggests some extralingual similarity as (unpleasant) tongue sensations.

6.1.2 Intralingual meaning

The sequences **akuppoi kedo aku ga nai* 'is harsh but doesn't have a harshness' and **aku ga aru kedo akuppoku nai* 'has a harshness but isn't harsh' are both adjudged as contradictory by the informant, indicating that the expressions are (cognitively) synonymous: the colloquial status of the suffix in AKUPPOI has been noted in 5.2, and some stylistic differentiation in terms of formality is likely.

As expected, the two items do not combine. In terms of affective value, based on combination patterns with evaluative taste terms, they are characterized as -OISHII. The inherently unpleasant nature of the taste quality involved is reflected in the fact that the food-substances concerned are normally subjected to a process of 'harshness extraction' (*akunuki*) during preparation, involving prolonged soaking in water.

6.2 The lexical system AJI III

AJI III contains five terms: KOOBASHII 'fragrant', DOROKUSAI 'muddy-flavoured', AOKUSAI 'grassy-flavoured', KUSAMI GA ARU 'strong-flavoured' and KAORI GA II 'aromatic'.

6.2.1 Extralingual meaning

AJI III terms are listed in the following chart together with lexical items denoting food-substances to which they are typically applied by the informant.

(1)

AJI III term	Focal exemplars	Taste norms
KOOBASHII 'fragrant'		HOOJICHA 'roasted tea' SEMBEI 'rice crackers'
DOROKUSAI 'muddy-flavoured'	KOI 'carp'	
AOKUSAI 'grassy-flavoured'	(URETE INAI) TOMATO '(unripe) tomatoes' AYU 'sweetfish'	
KUSAMI GA ARU 'strong-flavoured'	CHIIZU 'cheese' MATON 'mutton' KUSAYA 'dried mackerel' KUJIRA 'whale-meat'	

(1) *cont.*

	SARAMISOOSEEJI 'salami' MUGIGOHAN 'boiled rice mixed with barley' GAIMAI 'imported rice'	
KAORI GA II 'aromatic'	MITSUBA 'trefoil' MATSUTAKE '*matsutake*' (a type of mushroom)	KOOHII 'coffee' BANANA 'bananas' NORI 'laver' INDORINGO '*indoringo*' (a type of apple) HOOJICHA 'roasted tea' JASUMINTII 'jasmine tea' GEMMAICHA 'unhulled-rice tea'

It will be noted that one item, roasted tea (*hoojicha*), is cited against both KOOBASHII 'fragrant' and KAORI GA II 'aromatic'; this will be accounted for in the course of the analysis below. Also noticeable is the fact that, whereas KUSAMI GA ARU 'strong-flavoured' and KAORI GA II 'aromatic' are each associated with a considerable number of typical substances, the remaining terms are more restricted in this respect. This applies particularly to DOROKUSAI 'muddy-flavoured' and AOKUSAI 'grassy-flavoured', for which the substances given appear virtually to exhaust the practical possibilities. Once again, the dictionaries surveyed do not provide specific examples in the case of these terms; however, all relate KOOBASHII to the smell of dry-roasted or toasted foods and AOKUSAI to that of leaf vegetables (*aona*) when chopped.

By contrast with the terms of AJI I and AJI II, which range over the extralingual parameters of gustation and the tongue-sensations of pungency, astringency and harshness, AJI III terms relate to the olfactory parameter of taste. Morphologically, we have seen that all the terms incorporate forms which are semantically associated with the field of odour (cf. 5.2).[1]

6.2.2 Intralingual meaning

The same approach to the analysis of intralingual relations is adopted as for AJI I terms: response patterns to type-(B) questions are first examined, and this is followed by investigation of the possibilities of combination among the terms.

The semantic structure of AJI III proves to be much less complex than that of AJI I. First, as the presence of roasted tea (*hoojicha*) as a typical example for both terms suggests, KAORI GA II 'aromatic' and KOOBASHII 'fragrant' are related by hyponymy:

(B1)	Hoojicha wa *kaori ga ii*? ROASTED TEA/isolator/AROMA/ subject/GOOD, non-past 'Is roasted tea aromatic?'	Un, *koobashii* ne. YES/FRAGRANT, non-past/tag 'Yes, it's fragrant.'
(B2)	Indoringo wa *koobashii*? INDORINGO/isolator/FRAGRANT, non-past 'Are *indoringo* (apples) fragrant?'	Uun, *koobashiku* nai. *Kaori ga ii* kedo. NO/FRAGRANT/negative, non-past/AROMA/subject/ GOOD, non-past/BUT 'No, they're not. They're aromatic, though.'

The response patterns here indicate that KOOBASHII 'fragrant' functions as a hyponym of KAORI GA II 'aromatic'. As reflected in dictionary definitions, the informant reports that KOOBASHII is typically applied where the substance concerned has been dry-roasted.

The other three terms, DOROKUSAI 'muddy-flavoured', AOKUSAI 'grassy-flavoured' and KUSAMI GA ARU 'strong-flavoured', are also interrelated in terms of hyponymy:

(B3)	Koi wa *kusami ga aru*? CARP/isolator/SMELLINESS/ subject/BE PRESENT, non-past 'Is carp strong-flavoured?'	*Kusami ga aru* yo. *Dorokusai.* SMELLINESS/subject/BE PRESENT, non-past/ emphasizer/MUDDY-FLAVOURED, non-past 'Indeed it is. It's muddy-flavoured.'

(B4) Maton wa *dorokusai*?
'Is mutton muddy-flavoured?'

Uun, *dorokusaku* nai. Chotto *kusami ga aru* kedo.
'No, it isn't. It's somewhat strong-flavoured, though.'

(B5) Tomato wa *kusami ga aru*?
'Are tomatoes strong-flavoured?'

Un, chotto *aokusai* ne.
'Yes, they're somewhat grassy-flavoured.'

(B6) Chiizu wa *aokusai*?
'Is cheese grassy-flavoured?'

Uun, *aokusaku* nai kedo mono ni-yotte *kusami ga aru* ne. *Kaori ga ii* mono mo aru kedo.
NO/GRASSY-FLAVOURED/negative, non-past/BUT/THING/ACCORDING TO/SMELLINESS/subject/BE PRESENT, non-past/tag/AROMA/subject/GOOD, non-past/THING/integrator/BE PRESENT, non-past/BUT
'No, it's not, but some cheese is strong-flavoured. There are aromatic cheeses as well, though.'

where DOROKUSAI 'muddy-flavoured' and AOKUSAI 'grassy-flavoured' evidently function as co-hyponyms of KUSAMI GA ARU 'strong-flavoured'.

For other pairings of terms, bilateral simple negation is found:

(B7) Indoringo wa *kusami ga aru*?
'Are *indoringo* (apples) strong-flavoured?'

Uun, indoringo wa *kaori ga ii*.
'No, they're aromatic.'

(B8) Saramisooseeji wa *kaori ga ii*?
'Is salami aromatic?'

Uun, *kaori ga yoku* nai. *Kusami ga aru.*
'No, it isn't. It's strong-flavoured.'

(B9)	Hoojicha wa *kusami ga aru*? 'Is roasted tea strong-flavoured?'	Uun, hoojicha wa *koobashii*. 'No, it's fragrant.'
(B10)	Koi wa *koobashii*? 'Is carp fragrant?'	Uun, koi wa *koobashiku* nai. *Kusami ga aru.* 'No, it isn't. It's strong-flavoured.'
(B11)	Koi wa *aokusai*? 'Is carp grassy-flavoured?'	Uun, *dorokusai*. 'No, it's muddy-flavoured.'
(B12)	Ayu wa *dorokusai*? 'Is sweetfish muddy-flavoured?'	*Dorokusaku* nai. Ayu wa chotto *aokusai*. 'No, it isn't. Sweetfish is somewhat grassy-flavoured.'
(B13)	Indoringo wa *aokusai*? 'Are *indoringo* (apples) grassy-flavoured?'	Uun, *aokusaku* nai. Tottemo *kaori ga ii*. 'No, they're not. They're very aromatic.'
(B14)	Tomato wa *kaori ga ii*? 'Are tomatoes aromatic?'	Uun, *kaori ga yoku* nai. Chotto *aokusai*. 'No, they're not. They're somewhat grassy-flavoured.'

Bilateral responses for the remaining pairings (DOROKUSAI and KAORI GA II, DOROKUSAI and KOOBASHII, AOKUSAI and KOOBASHII) were in fact not elicited, the tendency being to respond at the more general level of the superordinate terms concerned. However, on the basis of the data presented, we may conclude that the intralingual structure of AJI III is as follows:

(2)

KAORI GA II 'aromatic'	KUSAMI GA ARU 'strong-flavoured'	
KOOBASHII 'fragrant'	AOKUSAI 'grassy-flavoured'	DOROKUSAI 'muddy-flavoured'

with horizontal lines separating superordinates (above) and hyponyms, and vertical lines separating contrasting terms.

Further light is thrown on the structure of the system when the possibilities of combination among the terms are considered. In the familiar matrix form (cf. 5.4.2.4), these can be represented as follows:

(3)

	KAORI GA II	KOOBASHII	KUSAMI GA ARU	AOKUSAI	DOROKUSAI
KAORI GA II 'aromatic'	//////				
KOOBASHII 'fragrant'		//////			
KUSAMI GA ARU 'strong-flavoured'			//////		
AOKUSAI 'grassy-flavoured'				//////	a-te d
DOROKUSAI 'muddy-flavoured'				d-te a	//////

As illustrated, combination among AJI III terms is highly restricted: only DOROKUSAI and AOKUSAI, the co-hyponyms of KUSAMI GA ARU, are syntagmatically compatible (in AND-combination). (Such a combination is in fact unlikely to find application in the real world.) The absence of the combinations KUSAMI GA ARU + DOROKUSAI, KUSAMI GA ARU + AOKUSAI, and KAORI GA II + KOOBASHII is to be expected from the analysis so far, since the terms of each pair are related by hyponymy. However, the syntagmatic incompability between the group consisting of KUSAMI GA ARU and its hyponyms, on the one hand, and the group consisting of KAORI GA II and its hyponym, on the other, cannot be explained on this basis.

Consideration of patterns of combination of AJI III terms with the evaluative taste terms reveals a clear correlation between incompatibility and affective

value. In terms of the symbolization introduced in 5.4.2.5, AJI III terms are characterized as follows:

(4)

KAORI GA II KOOBASHII	+OISHII
KUSAMI GA ARU DOROKUSAI AOKUSAI	–OISHII

Here, AJI III differs from AJI I in two ways. Firstly, all terms of AJI III are either +OISHII or –OISHII; that is to say, their affective characterization is clearly polarized, with no 'intermediate' cases of (–)OISHII. Secondly, there is a perfect correlation between OISHII-characterization and incompatibility. The terms of AJI III are thus split into two incompatible groups, the division correlating with a clear opposition of affective value:

(5)

<table>
<tr><td>KAORI GA II +
'aromatic'</td><td colspan="2">KUSAMI GA ARU –
'strong-flavoured'</td></tr>
<tr><td>KOOBASHII +
'fragrant'</td><td>AOKUSAI –
'grassy-flavoured'</td><td>DOROKUSAI –
'muddy-flavoured'</td></tr>
</table>

6.3 AJI: The overall field

Having discussed the semantic structure of the systems AJI I, AJI II and AJI III, we return at this point to consider briefly how the systems interrelate within the overall field.

In the first place AJI 'taste' functions as a head term for the whole field insofar as the members of AJI I, II and III occur in appropriate responses to the initial question-frame *X wa donna aji ga suru?* 'What kind of taste has X?'; it is precisely in view of this fact that we have referred to the whole field as the lexical field of AJI. However, there is evidence that the term AJI also functions, in certain contexts, in a more specific sense in which it relates to AJI I and AJI II but excludes AJI III.

The principal evidence comes from cases where the term AJI is explicitly contrasted with KAORI 'aroma', as in the following attested conversational example:

(6) Kono ringo wa *aji* mo ii shi *kaori* mo ii ne.
THIS/APPLE/isolator/TASTE/integrator/GOOD, non-past/AND/AROMA/integrator/GOOD, non-past/tag
'These apples have both a good taste (AJI) and a good aroma (KAORI), haven't they.'

where the inference from *aji (ga) ii* 'the taste is good' is that the apples exhibit the expected degree of sweetness or sourness (i.e. AJI I qualities) for the type of apple concerned. Similarly, the most likely inference from:

(7) Kono gureepufuruutsu wa *aji* ga warui.
THIS/GRAPEFRUIT/isolator/TASTE/subject/BAD, non-past
'This grapefruit has a bad taste (AJI).'

is that the grapefruit has a bitterness (i.e. NIGAMI GA ARU, again an AJI I term). AJI and KAORI are also contrasted in examples from the written secondary data:

(8) Kokuchishooyu ni kurabe, usukuchishooyu wa, *kaori*, *aji* no seibun tomo, oohaba ni osaerarete iru. (Koono, 1974: 69)
'Compared with full-strength soy sauce, mild-strength soy sauce is considerably restrained in both its aromatic (KAORI) and taste (AJI) components.'

(9) Koohii wa cha, toku ni ryokucha to kuraberu to, kurabemono ni naranai kurai no fukuzatsu na *kaori* to *aji* no yooso o motte iru. (Koono, 1974: 166–7)
'Compared with tea, especially green tea, coffee has incomparably more complex features of both aroma (KAORI) and taste (AJI).'

(10) Tatoe, nigai mono de atte mo, amami o kuwaeru to nigami ga kie, taberareru yoo ni naru. Tsumari, sono mono no motte iru *aji* o keshite shimau sayoo ga aru...Amami no kono sayoo wa, *aji* dake de wa nai. *Kaori* mo kesu. (Koono, 1977: 72)
'Even with a bitter substance, if one adds a sweetener the bitterness disappears and it becomes edible. In short, (sweetness) has the effect of

> erasing the basic taste (AJI) of the substance...This effect of sweeteners is not limited to taste (AJI). They also erase aroma (KAORI).'

Finally, the terms contrast in expressions involving the verb TSUKERU 'attach, apply' referring to the seasoning and flavouring of food: here, the use of the expressions *aji o tsukeru* 'add taste (AJI)' and *kaori o tsukeru* 'add aroma (KAORI)' relates clearly to the systems within whose domain the taste quality concerned belongs. Thus, with substances such as salt, soy sauce, *mirin*, etc., whose associated taste qualities fall within the range of AJI I, *aji o tsukeru* is used (*shio/shooyu/mirin de aji o tsukeru* 'season with salt/soy sauce/*mirin*).[2] On the other hand, with herbs such as trefoil (characterized as KAORI GA II 'aromatic'), we find *kaori o tsukeru* (*mitsuba de kaori o tsukeru* 'flavour with trefoil').

In these contexts the application of AJI relates to the AJI I parameters of gustation, pungency and astringency and, we may assume, to the AJI II parameter of harshness, given the acceptability of the collocations *aku ga aru aji/akuppoi aji* vs. **aku ga aru kaori/*akuppoi kaori*. However, it excludes the AJI III parameters of olfaction. This narrower application of AJI is made explicit in the following utterance produced by the informant:

(11) Ninjin o ninjintarashimeru no wa ninjin no *kaori* shika nai ne – shita ni kuru *aji* tte nai.
'What makes a carrot a carrot is just its aroma, isn't it – there is no taste that strikes the tongue.'

where the contrast is made between KAORI 'aroma' and *shita ni kuru aji* 'a taste (AJI) which strikes the tongue'.

This narrow sense of AJI may reasonably be glossed as 'tongue-taste'. The evidence indicates an organisation of the following kind, with the term AJI functioning at different levels of generality:

(12)

<table>
<tr><td colspan="2">AJI 'taste'
(AJI I, II, III)</td></tr>
<tr><td>AJI 'tongue-taste'
(AJI I, II)</td><td>KAORI 'aroma'
(AJI III)</td></tr>
</table>

6.4 Extended meanings

In this section we review extended meanings of AJI terms.[3] As well as the general structural perspective on semantic extension, we shall have in mind Williams' (1976) findings on synaesthetic semantic change in adjectives referred to in 2.3. Of particular interest in both connections is AMAI 'sweet': this term shows the widest range of extension, and for part of this range exhibits a clear semantic opposition with KARAI 'pungent'.[4] As we have mentioned in 5.4.2.4, the primary opposition amongst Japanese descriptive taste terms is undoubtedly felt to be between this pair. Thus, dictionaries of opposites (e.g. Nakamura, 1965: 28) list KARAI, rather than other taste terms, as the opposite of AMAI, and native speakers, when asked for the opposite of AMAI, overwhelmingly provide KARAI; this contrasts with the situation in Ponapean, for which Ishige cites MEM 'sweet' vs. KATIK 'bitter' as the basic opposition (1983: 18), and in English, where the primary opposition is intuitively SWEET vs. SOUR. It is important to stress that this status is not to be explained on the basis of the structure of AJI I alone: our analysis provides clear evidence for the opposed status of AMAI (and associated group A terms) within that system, but it does not point to KARAI rather than SUPPAI 'sour', NIGAI 'bitter' or SHIBUI 'astringent' as its binary opposite par excellence, since AMAI enters into relations of contrast and syntagmatic incompatibility equally with all these terms.

AMAI is extended in two broad directions, firstly to stimuli which appeal to other senses in a pleasurable way analogously to sweet tastes. Thus, it is applied to sweet smells, as of flowers: Morita (1989: 71) lists honeysuckle (*suikazura*), roses (*bara*), jasmine (*jasumin*), wisteria (*fuji*) and gardenia (*kuchinashi*) as examples. This synaesthetic extension is clearly in line with Williams' findings. More abstractly, it is also applied to sounds and communications with, in particular, a romantic appeal (cf. Field, 1990: 143–8): *amai merodii* is a soft, romantic melody, *amai koe* ('sweet voice') is a cajoling or seductive, usually female, voice, *amai ai no sasayaki* ('sweet whisperings of love') are 'sweet nothings', and *amai kotoba* ('sweet words') are honeyed or beguiling words. Synaesthetic extension from taste to sound also fits Williams' model, although we should note that the appeal in some of these examples pertains also to the content of the communication.

The other direction of extension of AMAI involves the notion of the lack of a particular quality relative to the norm (Field, 1990: 136). This extension begins in certain usages within the broad domain of taste itself, and is then carried into more abstract domains; opposition with KARAI holds here within taste and, partially, in more abstract uses.

Firstly, AMAI and KARAI function as antonyms within the domain of taste with reference to the dimensions of hotness and saltiness. The following are illustrative examples for hotness, where the English equivalent for AMAI is MILD:

(13) Kono karee wa sono karee yori *karai*.
THIS/CURRY/isolator/THAT/CURRY/THAN/PUNGENT, non-past
'This curry is hotter than that.'

(14) Sono karee wa kono karee yori *amai*.
'That curry is milder than this.'

(15) Kono karee wa ryoohootomo *karai* kedo kotchi no hoo ga mada amai.
THIS/CURRY/isolator/BOTH/PUNGENT, non-past/BUT/THIS/genitive/ALTERNATIVE/subject/STILL/SWEET, non-past
'Both these curries are hot, but, of the two, this is milder.'

(16) Kono karee wa toku-ni *amaku* mo *karaku* mo nai. Futsuu da ne.
THIS/CURRY/isolator/ESPECIALLY/SWEET/integrator/PUNGENT/integrator/negative, non-past/ORDINARY/copula, non-past/tag
'This curry isn't particularly mild nor hot. It's average.'

(13) and (14) each imply the other, and (15) and (16) are acceptable sentences, indicating that AMAI and KARAI here function as antonyms (cf. 4.4). The English equivalent for KARAI here, namely HOT, is also one of its glosses in AJI I. However there is a clear semantic distinction between the two cases: whereas in the basic field of descriptive taste terms the contrasting terms AMAI 'sweet' and KARAI 'pungent, hot' denote general taste qualities, here the antonymous pair AMAI 'mild' and KARAI 'hot' serve to place a given substance relative to the dimension of hotness. Thus, the associated question here is *Sono karee wa donogurai karai?* 'How hot is that curry?' (with KARAI as the unmarked term serving to indicate the dimension), not *Karee wa donna aji ga suru?* 'What kind of a taste does curry have?'. Clearly, all curry is KARAI 'hot' as far as taste quality is concerned, but a given curry may still be specified as AMAI 'mild' relative to the mild-hot dimension. The same antonymous usage occurs with reference to the dimension of saltiness. In this case the expression *shio ga amai* (SALT/subject/SWEET, non-past) tends to replace AMAI alone, as in the following example:

(17) Kono suupu wa shio ga *amai.*
THIS/SOUP/isolator/SALT/subject/SWEET, non-past
'This soup is (too) lightly salted.'

There is also a related compound noun here, AMAJIO (*ama-* 'sweet' + *shio* 'salt') 'lightly salted' (cf. *amajio no shake* 'lightly salted salmon').

This antonymy is extended outside the domain of taste, notably with reference to strictness of assessment (cf. Nishio, 1972: 329):

(18) Ano sensei wa kono sensei yori ten ga *karai.*
THAT/TEACHER/isolator/THIS/TEACHER/THAN/MARK/subject/ PUNGENT, non-past
'That teacher marks more strictly than this teacher.'

(19) Kono sensei wa ano sensei yori ten ga *amai.*
'This teacher marks more leniently than that teacher.'

(20) Kono sensei wa futaritomo ten ga *karai* kedo kotchi no hoo ga mada *amai.*
'Both these teachers mark strictly, but, of the two, this one marks more leniently.'

(21) Ano sensei wa toku-ni ten ga *amaku* mo *karaku* mo nai.
'That teacher doesn't mark particularly leniently or strictly.'

Here the adjectives most typically collocate with the noun TEN 'mark' as illustrated (SAITEN 'marking' and HYOOKA 'evaluation' are other possibilities), and the expressions are applied to teachers, judging panels, etc.

In all cases of AMAI/KARAI antonymy, we thus find AMAI located at the 'mild' end of the dimension. The uses within the taste domain clearly rest on an extralingual similarity with the basic system in the case of KARAI, in that reference is to dimensions of taste qualities (hotness, saltiness) which fall within its denotation in AJI I. As the natural contrast term within taste, AMAI is extended to serve as the antonym, denoting a relative lack of pungency rather than any presence of sweetness. The motivation for the subsequent extension to the domain of assessment is less transparent although, as indicated, MILD in particular participates in a similar analogy in English (cf. *mild/pungent criticism*).

AMAI, but not KARAI, is extended in a variety of further usages in this direction, often with a high degree of collocational restriction. Thus, AMAI is used in the sense of 'lenient' in collocation with such nouns as SHITSUKE 'discipline, upbringing' and nouns denoting persons who are the source of discipline such as OYA 'parent', SENSEI 'teacher'. In this more general sense of leniency, the antonym of AMAI is no longer KARAI but KIBISHII 'strict'; though the opposition with KARAI no longer holds, there is clearly an extralingual similarity with the above use of AMAI in relation to strictness of assessment. KIBISHII is also a possible antonym of AMAI in the expression *keibi ga amai* ('guarding is sweet') '(too) lightly guarded or protected' (of buildings, etc.), although KATAI 'hard (vs. soft), tight (vs. loose)' is also commonly used here. KATAI is likewise the antonym of AMAI in such expressions as *shubi ga amai* ('defence is sweet') 'defence is lax, vulnerable' (as in sports, etc.), *sen ga amai* ('stopper is sweet') 'cork/stopper (of a bottle, etc.) is not properly secured' and *neji no shimekata ga amai* ('tightening of screw is sweet') 'screw is not fully tightened', so that in the indicated usages of AMAI we find a chain of antonyms as follows:

(22)

Dimension	Antonyms
Pungency of taste Strictness of assessment	KARAI
Strictness of discipline	KIBISHII
Strictness of protection	KIBISHII/KATAI
Tightness of defence Tightness of securing	KATAI

AMAI also occurs in the expression *ha ga amai* ('blade is sweet') 'blade (of a knife, sword, etc.) doesn't cut properly', as in *Kono hoochoo wa ha ga amai* 'This (kitchen-)knife doesn't cut properly', where alternatives are *Kono hoochoo wa nibui* 'This knife is blunt' and *Kono hoochoo wa yoku kirenai* 'This knife doesn't cut well'. This usage has attracted attention from Williams (1976: 471), who interprets AMAI here as a touch term ('blunt') and notes that, as such, it constitutes an exception to the general rule for synaesthetic extension: while touch terms are often extended to the domain of taste, extension in the reverse direction is not generally attested (cf. 2.3). However, it is at least arguable that AMAI in this use is not to be characterized as a touch term in the strict sense: the semantic focus of *ha ga amai* appears to be on the effectiveness of the blade in the actual performance of its function, i.e. the process of cutting, rather than on its tactile physical features as such, clearly relevant though these are. Evidence for this comes from the fact that, based on mere physical inspection of the blade (as opposed to its cutting performance), the informant reports that she would use the evidential derivative (*Amasoo da na* '(The blade) looks as though it won't cut properly') rather than AMAI; this contrasts with the case of NIBUI 'blunt', which may be used on the basis of visual or tactile inspection (*Nibuku natteru* '((The knife, etc.) has become blunt') and thus qualifies more clearly as a touch term. Let us also note here the existence of the compound noun KIREAJI 'cutting quality', formed from the intransitive verb KIRERU 'cut' plus AJI 'taste'; this noun typically collocates with the general evaluative adjectives II 'good'/ WARUI 'bad' (*Kono hoochoo wa kireaji ga ii* 'This knife cuts well'), but the collocation *kireaji ga amai* ('cutting-taste is sweet') 'doesn't cut well' also occurs. Katsumata (1954: 22) glosses this usage of AMAI as 'not well-tempered; soft' and provides the Japanese paraphrase *kitae ga tarinai* 'tempering is insufficient', with explicit reference to the lack of an expected property.

As a final extension of AMAI in this direction we find usages referring to a lack of realism or hard-headedness in attitudes towards the practical world. Common collocations here are with nouns such as KANGAE 'thinking', KANGAEKATA 'way of thinking', YOMI 'interpretation' and, in adverbial form, with verbs such as MIRU 'look at, view'. Thus, *kangae ga amai* ('thinking is sweet'), applied to persons, may be glossed as 'be overly optimistic, insufficiently realistic in one's thinking' and *amaku miru* ('view sweetly') as 'underestimate (an opponent, etc.), take a simplistic, over-optimistic view of (a problem, etc.)'. Similarly, the expression *ningen ga amai* ('person is sweet') is used as a personality term: *Ano hito wa ningen ga amai* 'He/she is insufficiently realistic, soft-headed (in his/her view of the world)'.

To sum up the extended meanings of AMAI, these divide into two broad directions, as follows:

(23)

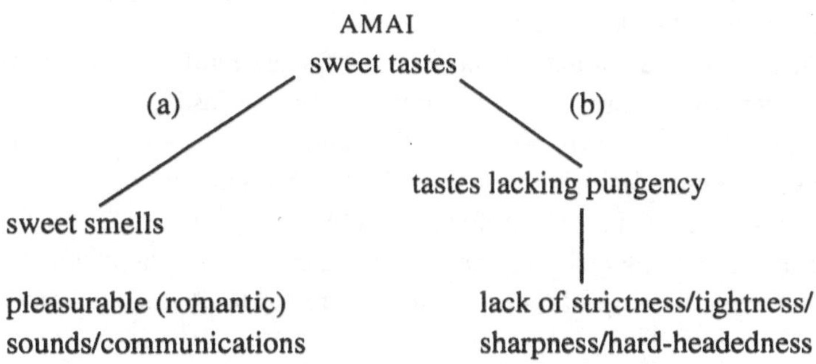

(a) involves extension to smells with obvious perceived similarity to the basic taste quality, and to romantic sounds and communications. In (b) we find AMAI extended, firstly within the broad domain of taste, as the natural antonym of KARAI and denoting a relative lack of pungency, be this hotness or saltiness. Further extension occurs, partially in company with KARAI, to more abstract domains where reference is to a lack of strength, or to a slackness, of various kinds. A notable feature of these extensions is that AMAI here denotes evaluatively neutral or, in many cases, unfavourable qualities, in contrast with its positive status in the basic field, and in this respect it behaves quite differently from English SWEET. These (b) extensions provide a good illustration of semantic transfer mediated by intralingual relations (i.e. by opposition to KARAI) rather than by perceived extralingual similarity with the basic meaning (cf. Lehrer, 1978: 117).

Turning to KARAI, we have already seen the extension of this term to strictness of assessment. In addition, there is a further application of the AMAI/KARAI opposition within the taste domain: strictly speaking, this involves compounds rather than the independent adjectives, but these items are of interest in that they illustrate an extension of the meaning of KARAI, rather than of AMAI as in the examples above. The first pair of compounds, AMAKUCHI and KARAKUCHI (*ama-/kara-* + *kuchi* 'mouth') are typically applied to types of *sake* or wine and correspond to English 'sweet' and 'dry', respectively. Denotationally, the distinction here is between sweetness (AMAKUCHI) and non-sweetness (KARAKUCHI), rather than between sweetness and pungency as in the basic field: in terms of taste quality at a general level, dry beverages will often be characterized as bitter, sour, or astringent rather than as pungent. The second pair AMATOO and KARATOO (*ama-/kara-* + *too* 'group, party'), may be glossed as 'a person who likes sweet things' and 'a person who enjoys alcohol', respectively, these being regarded as basic categories into which individuals can be grouped by virtue of their dietary preferences. Once again, the extralingual opposition is one of sweetness vs. non-sweetness, rather than of sweetness vs. pungency, and in both cases the meaning of KARAI is extended to produce an opposite (in this case, a complementary) of AMAI 'sweet'.

There is thus an ample linguistic basis for the intuition that AMAI/KARAI constitutes the primary opposition among Japanese descriptive taste terms. We have seen that any such opposition will naturally involve AMAI, given its special status in the basic field, but why should KARAI, rather than some other term, be chosen in the case of the present examples? It is reasonable to look to the cultural salience of the taste qualities concerned as a major factor here: among the non-sweet tastes, the range of taste qualities covered by KARAI includes those of salt, soy sauce and *miso* (fermented soybean paste), whose central position as seasonings in the Japanese culinary tradition has already been pointed out (cf. 4.6). Indeed, the presence in Japanese of a single, morphologically simple taste term ranging over these qualities contrasts markedly with the situation in English, for example, where corresponding terms are either morphologically complex (SALTY) or extended from other domains (HOT, PUNGENT, PIQUANT; cf. Williams, 1976: 475–6).

Of the remaining terms, AMAZUPPAI, 'sweet-sour' and SUPPAI 'sour' may both be used to describe smells, in what we have seen to be a regular synaesthetic transfer. Thus, AMAZUPPAI may be used to describe the smell of certain flowers (cf. Morita, 1989: 71); SUPPAI is used, for example, of vinegary cooking smells, and Nishio (1972: 99) lists attested literary examples referring to the sour smell of *sake* (*suppai sake no nioi*) and to the smell of excessive perspiration (*suppai hodo asekusaku natte shimau* 'become so sweaty-smelling as to be sour'). Other AJI I terms (like evaluative taste terms in Japanese (cf. 4.3)) do not appear to be used directly of smells: Nishio (1972: 108) cites an example containing the collocation *karasoo-na nioi* 'pungent-seeming smell', with the evidential derivative rather than the basic term, referring to the smell of curry powder. From AJI III, *aokusai nioi* 'grassy-smelling smell' is attested by Nishio (1972: 107), in examples referring to the smell of grasses and carbolic acid. This item incidentally points to a potential problem for Williams' model: morphological evidence (*ao-* 'blue, green' + *kusai* 'smelly') suggests that AOKUSAI is in origin a smell term yet, as our analysis has shown, it qualifies semantically as a taste term in the modern language and, as Lehrer points out (1978: 119), transfer from smell to olfactory components of taste seems a natural extension. 'Taste' for Williams must be interpreted in a narrow sense (as 'tongue-taste': cf.6.3) if his scheme is to be saved from exceptions of this kind.

Beyond synaesthetic extensions, SUPPAI 'sour' also occurs in the idioms *kuchi ga suppaku naru* ('one's mouth becomes sour')/*kuchi o suppaku suru* ('make one's mouth sour'), applied to situations where one imparts instructions, advice, etc. repeatedly but to no avail: *kuchi o suppaku shite iu* 'say/tell (someone) time and time again'. (The archaic SUI 'sour', but not SUPPAI, also occurs in the idioms *sui mo amai mo kamiwakeru/shitte iru* ('savour/know both the sour and the sweet') 'experience/know both sides of life', which exhibit an opposition between AMAI and SUI.) AMAZUPPAI 'sweet-sour' is applied to experiences, of a pleasant but poignant nature: *amazuppai hatsukoi no aji* (Hayashi et al., 1987: 15) 'the sweet-sour taste of first love', which also involves an extension of AJI 'taste'. NIGAI 'bitter', as we have seen, denotes what is commonly regarded as the unpleasant taste par excellence, and this aspect of its meaning is clearly preserved in expressions such as *nigai keiken* 'a bitter experience'; close to NIGAI in this usage is the adjective TSURAI 'trying, painful', which has a much wider range of application. *Nigai kao (o suru)* '(make a) bitter face' refers to a displeased facial expression, like that of someone who has tasted a bitter substance;[5] cf. also *nigawarai (o suru)* '(give a) bitter/grim smile', a compound of NIGAI and the verb WARAU 'laugh, smile'. HORONIGAI 'pleasantly bitter' is likewise extended to experiences in collocations such as *horonigai jinsei* ('pleasantly bitter life') 'a life which has its ups and downs, but is basically rewarding': its meaning here is somewhat similar to that of the English term BITTERSWEET, whose use in the modern language appears to be restricted to extended contexts of this kind. Clearly, the affective difference between HORONIGAI and NIGAI in the basic field is maintained in these usages.

Like NIGAI, SHIBUI 'astringent' denotes an unpleasant taste quality, and the term is similarly applied to facial expression in *shibui kao (o suru)* '(make an) astringent face, (make an) unwilling, reluctant face'. Negative evaluation is also apparent in the use of the term applied to persons with the meaning 'stingy'. At the same time, however, SHIBUI is widely used as a term of aesthetic approval, where it may be glossed as 'austere, tasteful'; in this sense it is a key term in Japanese aesthetics, contrasting with terms such as HADE-NA 'gaudy, loud' and denoting a restrained but robust quality that is a prized feature of much Japanese artistic tradition. Its range in this usage is wide: typical collocations are with nouns such as IRO 'colour', GARA 'pattern', BUNSHOO 'writing, written style', KOE 'voice', and the noun KONOMI '(aesthetic, etc.) taste' itself. In this case, the semantic link with the taste term appears tenuous by comparison with the other examples considered in this section, and treatment as homonymy may be more appropriate here. Apropos of this use, *shibu* (the astringent extract of persimmons) is used in the production of a traditional type of paper admired for its tough but elegant qualities (Kawakita, 1960–1: 33), and historically it is possible that this noun, rather than the taste adjective, has been the basis for the extension.

These exhaust the common extensions of AJI I terms. Of the terms of AJI II and AJI III, DOROKUSAI 'muddy-flavoured' is applied to persons, actions, etc. with the derogatory meaning of 'countrified, crude' (cf. Seidensticker and Matsumoto, 1982: 617). *Aku ga tsuyoi* ('harshness is strong': cf. the AJI II terms AKU GA ARU 'has a harshness'/AKUPPOI 'harsh') is used as a personality term, likewise derogatory, meaning 'disagreeably forceful'.

Finally, what of the term AJI 'taste' itself? As we have seen in 6.3, AJI is used within the taste domain with the specialized sense of 'seasoning', i.e. with reference to the imparted taste-component of cooked dishes; in addition to *aji o tsukeru* 'add taste', *aji ga usui* ('taste is thin') 'lightly seasoned' and *aji ga koi* ('taste is thick/concentrated') 'heavily seasoned, rich-tasting' are common collocations here. As for more abstract domains, unlike TASTE in English AJI is not used in the sense of a person's aesthetic or other preferences, where Japanese employs the terms KONOMI or SHUMI. However, it occurs in extended uses such as the following, where it is commonly paralleled by English terms such as TASTE, FLAVOUR or SPICE: thus, *hatsukoi no aji* 'the taste of first love', *Nihon no ongaku no aji* 'the spirit/flavour of Japanese music', *aji ga aru hanashi* 'story having spice/interest', *aji ga nai e* ('picture which has no taste') 'an insipid/lifeless painting'.

7 *Retrospect and prospects*

In this concluding chapter we relate the findings of our study of Japanese taste terms to some of the general questions raised earlier concerning taste and taste vocabularies (7.1). In 7.2 we consider some directions in which the work of this study might be extended.

7.1 Review of findings

Firstly, the linguistic evidence from Japanese speaks clearly in favour of those psychologists and others who have protested that gustatory qualities alone constitute an unduly narrow framework within which to study the workings of the sense of taste in everyday life (cf.1.1). The lexical semantic structure of Japanese does not support the primacy of the four 'basic' tastes: rather, this study has shown that the domain of the field of AJI in Japanese ranges over the parameters of gustation, the tongue sensations of pungency, astringency and harshness, and olfaction; while linguistic subdivisions are made within this range, it is clear that the gustatory qualities do not constitute a separate linguistic domain, but fall together with pungency and astringency under AJI I. If a division can be established, it is rather between 'tongue-tastes' (covered by AJI I and AJI II, and subsumed under the range of AJI in the narrower sense) and olfactory qualities (AJI III) (cf. 6.3). At the same time it is important to note that other dimensions of the taste experience in the broadest sense, such as texture, consistency, shape and temperature, are clearly excluded from the domain of AJI.

What conclusions do our findings suggest with respect to the general points of potential difference between taste and colour vocabularies raised in 1.3 under the headings of complexity, poverty, cultural channelling, affectivity and individual variation? The first of these concerns the relative complexity of the sense of taste in terms of the variety of receptors and parameters involved. Insofar as we have found a basic field divided into three separate systems in place of the monolithic field normally characteristic of colour terms, we may say that this complexity is indeed reflected linguistically in the structure of the field of AJI. However, this question is clearly bound up with the breadth of our definition of 'taste': as we have seen in 6.3, AJI has a narrower sense ('tongue-taste') in which its range excludes AJI III and where complexity is correspondingly reduced. Linguistically, it appears that a narrower definition may be justified insofar as, historically at least, the primary application of AJI III terms seems to have been in the domain of smell (cf. 6.4).

The commonly held view that the vocabulary of the proximate senses is 'poorer' than that of the distal senses is likewise somewhat difficult to assess. We have noted that this must be interpreted as referring to basic terms, since the vocabulary in this area is readily expanded by means of morphological and semantic extension where the need is felt, as in the description of wine in English. Even in this sense, it is almost certainly incorrect if interpreted on a purely quantitative basis since, as we know from the work of Berlin and Kay and subsequent studies, there are languages which possess only two, or three, basic colour terms. More relevant to this question is no doubt the fact that, unlike colour terms, taste terms do not constitute a lexical continuum, insofar as there are taste qualities which fall completely outside the range of the set of taste terms (cf. (A31) (5.1), repeated here):

(A31)	Shake wa donna aji ga suru?	Shake no aji to shika ienai-daroo ne.
	'What kind of a taste does salmon have?'	'I suppose you could only say it tastes like salmon.'

If 'poverty' is taken in this sense, then the evidence from Japanese supports the view: in describing taste, one is often reduced to talking in terms of specific 'things'. At the same time, we should not lose sight of the common-sense principle that the resources of a language are in general adequate to the everyday needs of its speakers; in accordance with this principle, one must assume that taste vocabularies, like those of any other domain, meet the everyday requirements for linguistic encoding efficiently enough.

Culture-specific aspects of the input to the sense of taste are reflected most clearly in the 'exotic' nature of certain of the substances associated with the various terms as focal exemplars in axioms (2.2) and in taste norms (1.4), both of which, we have argued, occupy an important place in semantic investigations of taste vocabularies. We have also suggested a cultural basis, namely the prominence in traditional Japanese cuisine of a particular range of seasonings, for the salience among taste terms of the AMAI vs. KARAI opposition (cf. 6.4). Despite the universal or widespread availability of certain basic substances, such as breast-milk, salt and honey, there can be little doubt that man's dietary world is inherently subject to greater variation than fundamental aspects of the physical world (blood/fire, the sky, leaves/grass, etc.) which we may assume provide the anchoring points for the extralingual meaning of basic colour terms.

AJI terms also reflect the affective aspect of the sense of taste, most obviously in the different patterns of combination exhibited with evaluative taste terms (cf. 5.4.2.5). In turn these differences correlate, more or less precisely, with the relationships of syntagmatic incompatibility amongst individual terms, with group A terms centring on AMAI 'sweet' occupying a particularly distinctive position in this respect (cf. 5.4.2.4). Overall, the linguistic evidence clearly indicates that taste sensations differ in affective appeal according to their type and degree, and this has no general parallel in the field of colour.

Finally, we have noted that, intuitively, taste descriptions and judgements exhibit a greater degree of individual variation than we would expect to find in colour; this is no doubt connected with the affective aspects of taste, and finds its most obvious manifestation in the application of evaluative terms to given substances. While this can only be empirically confirmed through comparative studies of the usage of different speakers, certain of our findings are indicative in this respect. Thus we have seen, for example, that both NIGAI 'bitter' and HORONIGAI 'pleasantly bitter' – terms which differ markedly in affective value – are commonly applied to beer, according to the preferences of the individual concerned (5.4.1). More generally, we have seen that evaluative taste terms fall squarely, and descriptive taste terms marginally, within the category of subjective adjectives in Japanese (4.6, 5.4.1). With these adjectives, qualities are presented as products of an interaction between a stimulus and an experiencer rather than as independent properties of entities, and on this continuum evaluative taste terms, in particular, stand clearly opposed to colour terms, falling together rather with adjectives from the perceptual domains of pain and temperature.

7.2 Future directions

This study has examined the semantic structure of the central taste vocabulary of one language. As we have seen, linguists have hitherto paid little attention to taste terms, and further work is needed in many directions in this area, both in Japanese and in other languages.

Within Japanese, firstly, there is a need for similar studies of other idiolects of the variety. In general, some variation in linguistic repertoire is to be expected with respect to more peripheral members of lexical fields (cf. 3.2): we have seen clear evidence of this in the present field especially with AJI II terms, and have also noted the apparent availability of variant -MI GA ARU terms relating to sourness, originating in dialectal differences (6.1, 5.2). It seems reasonable to expect that women will typically possess a fuller active vocabulary in this particular field, so that sex may well be one important social parameter here. Studies should also investigate variation in the application or understanding of these peripheral terms.[1] Secondly, the range of this study may usefully be extended to other domains of the Japanese vocabulary of taste (in the widest sense), using collocation with evaluative taste terms as a criterion for delimitation as indicated in 4.6. Here belong items relating to various aspects of 'taste-feel' where, we have suggested, Japanese possesses a variety of characteristic terms (1.4). As one illustration of the rich lexical resources available, we may cite the pair of terms SHIKOSHIKO-SHITE IRU and KORIKORI-SHITE IRU. Both are positive affect terms denoting resilience to the bite: of the two, SHIKOSHIKO-SHITE IRU describes a lower degree of firmness, being associated (in taste norms) with noodle-type foods where its closest English equivalent is '*al dente*', whereas KORIKORI-SHITE IRU denotes a higher resilience associated with such items as jellyfish and certain pickles. As these examples illustrate, many terms referring to tactile dimensions of taste belong to the mimetic stratum of the vocabulary. Finally, more attention should be directed to the question raised by Kuipers (1984) concerning the uses to which taste terms are put in actual discourse. As far as everyday conversation is concerned our impression is that, while evaluative taste terms, particularly OISHII, are frequently used in Japanese, the primary use of descriptive taste terms, outside explanations of unfamiliar food items, is with reference to particular food specimens where they likewise carry evaluative – typically, negative – implications as discussed in 1.4. This doubtless reflects the importance of affectivity in taste, and the existence of taste norms as part of shared background knowledge in the culture concerned. Kuipers points to a similar situation in Weyéwa when he stresses 'how unusual it is, in ordinary discourse, to use [descriptive] taste words at all' (1984: 92): aside from the extended use of the bitter/bland opposition in ritual speech referred to in 1.2, descriptive taste terms in Weyéwa almost always occur modified by (often emotionally charged) intensifiers such as 'too'/'not...enough'/'really'/'rather' (1984: 93).

More generally, there is clearly a need for detailed semantic studies of the taste vocabularies of other languages, to provide data for cross-linguistic and cross-cultural comparison which, in turn, may be expected to contribute to scientific understanding of the sense of taste (cf. 1.2). Important cross-linguistic topics here include the delimitation of taste, basic taste terms, and semantic extension.

The general question of the delimitation of the taste domain, including the important role of question-and-answer procedures, has been discussed in 3.1 and is clearly fundamental to cross-linguistic comparative work. Thus, the everyday range of application of English words such as TASTE (noun and verb) and FLAVOUR and their counterparts in other languages is a basic topic for investigations in this area. Evidence from both Japanese and English demonstrates a wider range than that defined for the noun TASTE in scientific language (i.e. gustation). In the case of the Japanese term AJI we have found grounds for distinguishing a narrower and a broader range (this latter including olfaction) (6.4), but both exclude various 'taste-feel' qualities: SHIKOSHIKO-SHITE IRU '*al dente*' and KORIKORI-SHITE IRU 'resilient' are not AJI terms on our analysis. For this 'super-domain', embracing a wide variety of qualities perceived during ingestion, we have pointed out that co-occurrence with evaluative taste terms (OISHII/UMAI/MAZUI) provides a delimitative criterion (4.6). Finally, question-and-answer procedures have served to delimit an additional important domain within AJI 'tongue-taste', namely that of the system AJI I (5.3). We have thus effectively distinguished the following linguistic domains for this area in Japanese:

(1)

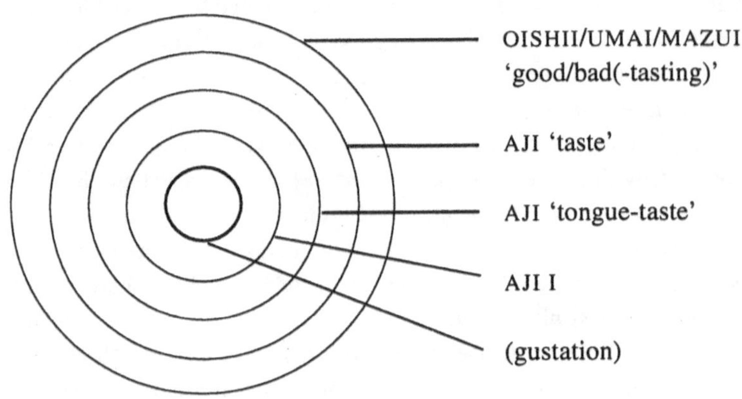

The disposition of basic taste terms across languages is a natural topic for investigation in the light of familiar studies in the field of colour: given the work of Berlin and Kay and their successors, it is tempting to entertain the hypothesis that basic taste vocabularies exhibit comparable universal principles of organization and development. While any suggestions in this direction can only be highly speculative in the present state of research, we first consider here the issue of basic taste terms with reference to modern Japanese, and then survey available evidence from further afield with a view to identifying potential cross-linguistic tendencies.

Berlin and Kay employ four principal criteria for identifying basic terms in the field of colour (1969: 5–7). These terms should be (a) monolexemic, i.e. non-transparent in composition, (b) not included semantically under a higher term, (c) applicable to a wide range of objects, and (d) psychologically salient, implying early occurrence on elicited lists, stability of reference across informants and across occasions of use, and occurrence in all idiolects. In applying these criteria to Japanese taste terms, we naturally need to confront the problem of domains just illustrated: which of the various delimited domains are we to select as representative? For present purposes we shall elect to focus on the domain of AJI I, as the 'tightest' domain incorporating what are undoubtedly felt to be central taste qualities such as sweetness, and as the area for which cross-linguistic data is primarily available (though often restricted to gustatory qualities, as seen in 1.2).[2] We shall refer to it here as the domain of tongue-taste.

Applied to AJI I terms, the criteria yield the following results:

Basic:	AMAI 'sweet'
	KARAI 'pungent'
	NIGAI 'bitter'
	SHIBUI 'astringent'
Borderline:	SUPPAI 'sour'
Non-basic:	AMAZUPPAI 'sweet-sour'/AMAKARAI 'sweet-pungent'/
	HORONIGAI 'pleasantly bitter' (not monolexemic)
	SHIOKARAI 'salt' (not monolexemic, hyponym of KARAI)
	SHOPPAI 'salty' (restricted to colloquial contexts)

The main question-mark over SUPPAI concerns its stylistic status: as we have seen (5.2), SUPPAI has historically superseded SUI in Tokyo and adjoining areas of eastern Japan, but geminate consonants tend to carry colloquial overtones in the native vocabulary and some speakers would avoid using SUPPAI in the written language, preferring SUI or circumlocutions such as *sammi ga tsuyoi* 'sourness is strong'. However, this restriction is less strong than with SHOPPAI 'salty';[3] moreover, SUI is felt as archaic by many speakers. On balance, it appears preferable to admit SUPPAI to basic term status, giving a total of five basic taste terms for modern Japanese;[4] from a cross-linguistic perspective, the presence of a term ranging over hot and salty tastes (KARAI) and of a basic term denoting astringency (SHIBUI) are striking features.

The next requirement is to adduce comparable data, i.e. data resulting from similar procedures of delimitation and of identification of basic terms, from other languages and from different historical stages of languages. Little, if any, such data is currently available; in its absence we advance some tentative suggestions based on existing information on taste vocabularies, including the studies mentioned in 1.2 and some additional lexicographical and other sources.

Firstly, existing evidence suggests that all languages possess at least two basic terms within the domain of tongue-taste, one having a denotational range which includes sweetness, the other taking in disagreeable taste qualities. Myers (1904: 122) cites two such languages, one from Uganda and the other from Malekula in the New Hebrides: he describes the Ugandan pair KUWOMA/KAWA as being applied to sweet, salt[5] and other agreeable flavours, including that of 'certain sour fruits liked by the Baganda' (KUWOMA), and to unpleasant flavours including both sourness and bitterness (KAWA), respectively (1904: 122, 124). (Myers also mentions New Guinea, the New Hebrides and much of Polynesia as having a common word denoting salt, sour, and bitter tastes, though without referring explicitly to two-term systems in these cases (1904: 124).) Such systems appear to be organized around the affective aspects of tongue-taste, with one term denoting sweetness but also pleasantly salt and pleasantly sour tastes, the other denoting unpleasant qualities including bitterness and extreme sourness, and perhaps also extreme saltiness together with pungency/hotness and astringency (these latter qualities, let us recall once more, being largely ignored at the expense of gustation by early investigators such as Myers). Beyond this two-term stage we may hypothesize progressive differentiation, with the emergence of additional terms relating to non-sweet tongue-taste qualities. Thus the Todas appear to have distinct basic terms for sweet, bitter and sour/salty (cf. Rivers, 1905: 388), and the conflation of sour and salty tastes under a single term in various languages is also referred to by Myers (1904: 119, 120, 124). Myers also alludes to bitter/salty, bitter/astringent, and bitter/pungent conflations (1904: 124–5), and our Japanese data have illustrated that salty and hot may fall together within the range of a single term. Only systematic studies of a range of languages will determine the precise nature of the conflations found and the presence, if any, of universal tendencies in this respect.[6] Further elaboration of tongue-taste vocabularies may take us beyond basic terms (as with English SALTY); on the evidence of modern Japanese, expected developments here will include the emergence of terms encoding differences of degree and/or affectivity (cf. HORONIGAI 'pleasantly bitter', SHOPPAI 'salty', PIRITTO(-SHITE IRU) 'piquant')[7] and taste combinations or blends (cf. AMAZUPPAI 'sweet-sour', AMAKARAI 'sweet-pungent').[8]

A complication not considered in the discussion above concerns the place of evaluative taste terms in relation to basic tongue-taste terms. In Japanese it is clear that these are quite distinct systems with totally different ranges, and we also find distinct evaluative taste terms co-existing with apparently basic tongue-taste terms such as MEM 'sweet' and KATIK 'bitter' in the case of Ponapean, for which Ishige cites IOU 'good-tasting' and also indicates the presence of an opposite term (1983: 18). In some systems, however, the functions of descriptive and evaluative taste terms appear to be conflated, the presence of a single term rendered as both 'sweet' and 'good-tasting', in particular, being widespread (cf. Myers, 1904: 119; Ishige 1983: 21, with reference to Anatolian Turkish); thus, the New Guinea language Fore appears to possess two taste terms, IGÁ̲WE and ASA̲BAYÉ, glossed by Scott as 'sweet, tasty, pleasant, delectable' and 'be bitter, be sour', respectively (1980: 36, 25) and in Kwamera, an Austronesian language of Vanuatu, ɨSIÉN glossed as 'savoury, tasty, properly sweet, properly salty, well-seasoned' appears to contrast with ɨFIA 'bitter, sour, too salty, bad tasting, unsavoury' (Lindstrom, 1986: 46, 51). It is certainly conceivable that some taste-term systems are general in this respect, and that the differentiation of taste evaluators and tongue-taste terms represents a later development. Once again, this is a question which must await careful descriptive study, with particular attention to the identification of the domains involved. A related problem concerns the status of terms such as 'tasteless' and 'bland', which are attested quite widely, as in Galelan (Ishige, 1983) and Weyéwa (Kuipers, 1984). In principle such terms appear capable of taking their place in descriptive systems (cf. COLOURLESS in relation to English colour terms) or evaluative systems (cf. English INSIPID): thus, Galelan KAWA is grouped as a negative term among evaluators (Ishige, 1983: 19), while Weyéwa KÓBA is listed as a basic term alongside the tongue-taste terms sour, sweet, salty, bitter, tart and pungent (Kuipers, 1984: 92) and its status as a descriptor (which may of course carry an affective component of meaning) is supported by its opposition with PÓDDU 'bitter' in ritual speech. Two terms of this type are reported for Anatolian Turkish: one (TATSIZ) stands opposed to TATLI 'good-tasting/sweet' and is glossed as 'tasteless, bad-tasting', the other (YAVAN) as 'not good-tasting' with reference to lack of sweetness, of saltiness or of oiliness (Ishige, 1983: 21). Kwamera, cited above as another language with an apparently generalized system, also has two terms: -MɨMIS 'bland (as water), tasteless (in a positive sense)' and -SIU 'bland, unseasoned (as food which ought to have flavour)' (Lindstrom, 1986: 77, 124). *Prima facie* the Kwamera terms would appear to be a descriptor and an evaluator, respectively; while this is conjectural in the absence of detailed study, it suggests the existence of intermediate types of systems which combine general (i.e. conflated descriptive/evaluative) terms with differentiated terms in different areas of the overall domain. That is to say, we may expect to find languages of the following three types with respect to the structure of these areas of the taste vocabulary:

(2)

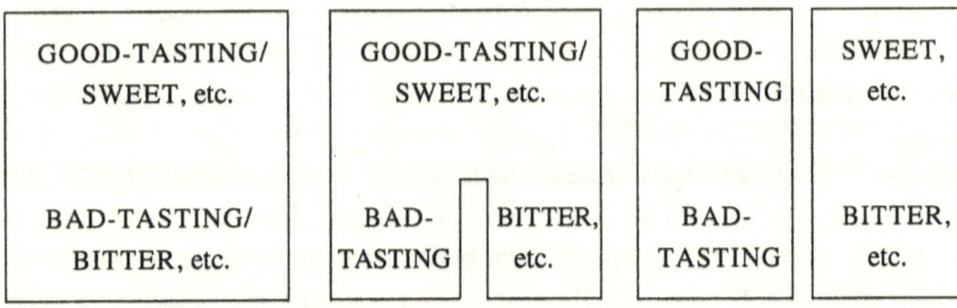

I. Generalized system | II. Intermediate system | III. Differentiated systems

On the evidence presented above, Fore appears to belong to type I, Ponapean (like Japanese) to type III. Kwamera may belong to type II in distinguishing descriptive and evaluative terms denoting lack of taste; more generally, the particular conflation of 'sweet' and 'good-tasting' appears to be attested in several languages, as noted earlier.

Evaluative taste terms in themselves raise several interesting questions for cross-linguistic investigation. Irrespective of the various possibilities of interaction with tongue-taste qualities explored above, let us assume that all languages provide lexical items for the expression of general appraisal of food substances. Languages differ in the degree of specialization of basic terms to this task: thus Japanese OISHII is specifically a taste evaluator, unlike English GOOD, NICE, etc. which are general evaluators and require syntagmatic supplementation for greater specificity (cf. 4.1, 4.7).[9] Japanese UMAI and MAZUI are arguably somewhat more general than OISHII (cf. 4.7), but still contrast with terms such as II 'good'.[10] In connection with taste evaluators we have also noted a tendency to operate in many contexts in Japanese with the positive term and its negation (*oishii/oishiku nai*) rather than with the antonymous pair (OISHII/MAZUI), even to the extent of pressing the negative into service as a complementary opposite (cf. 4.5). Similar usage is reported for other languages (e.g. Ponapean: cf. Chapter 4, note 12), and the semantics and pragmatics of these oppositions would doubtless repay further study.[11]

As a final cross-linguistic topic, semantic extension involving taste terms has been explored from a historical perspective and within the confines of synaesthetic transfer by Williams (1976) who, as we have seen (2.3), has identified certain universal tendencies in the directions of such extensions. For English, Lehrer (1978) surveys extensions of terms which occur in the domain of wine-tasting, including taste terms in the narrower sense. Here, as elsewhere, we need to be mindful of the domain of 'taste' to be examined: we shall focus on tongue-taste terms, although evaluative taste terms, for example, also merit study from this viewpoint. We first discuss extensions of tongue-taste terms into other domains; secondly, and more briefly, we consider the question of extension into the domain of tongue-taste.

What domains figure in taste-term extensions? In synaesthetic transfers, we have seen that Williams proposes that taste terms are in general extended directly only to smell or to sound; via sound, they may then also transfer indirectly to colour, but transfer to the domains of touch or dimension is prohibited (1976: 463–6). Williams' scheme is thus clearly testable, and in general holds up well in both English and Japanese: putative exceptions tend either not to be established usages, or to involve borderline cases of the domain concerned and thus to highlight ever-present questions of delimitation (cf. 6.4, also Williams, 1976: 468 on the status of FAINT as a smell term). Outside synaesthesia, both Lehrer (1978: 98–9) and Dirven (in an account of the extensions of SWEET; 1985: 104–6) point to extensions within the taste domain itself, as in *sweet (vs. salt) water*, and we have seen comparable examples for AMAI and KARAI in Japanese (6.4). Further afield, prominent domains noted in these studies are non-sensory experience (*sweet dreams*, *bitter experience*) and personality or personal evaluation (*sweet person*, *sweet child*). This latter domain falls within Dixon's 'human propensity' semantic type, one of seven such types associated with the word-class of adjectives in English (1982: 15–16). As mentioned in 1.3, descriptive taste terms basically belong to Dixon's 'physical property' type, but he notes that derived adverbs, where they exist, typically correspond to 'human propensity' type extended meanings of the adjectives, and he cites collocations such as *sweetly request*, *sourly reject*, *bitterly protest* and *bitterly jealous* (1982: 31). BITTER, in particular, seems to be prominent in this area, as illustrated by additional common collocations such as *bitterly resent*, *bitterly opposed/divided*, *bitterly fought*, *bitter enmity/rivalry*, etc. Extensions to non-sensory experience and to human propensity were both attested for Japanese (cf. 6.4); this latter area covers much semantic ground, and more detailed studies of extension will need to distinguish smaller domains within it. From other languages, we have noted the extension in Weyéwa of the terms 'bitter' and 'bland'. These are applied to the ritual status as 'prohibited' or 'permitted', respectively, of women, animals, or rice fields, and this is interpreted by Kuipers as expressing an analogy between gustatory and moral evaluation, mediated by the notion of appropriacy for consumption (1984: 94–6): under certain ritual conditions these items are not available (for marriage or consumption), just as bitter tastes are inappropriate for ingestion. As a final example, we may mention the case of Dyirbal GULI, reported by Dixon to have the senses 'sharp-edged' and 'sweet',[12] with antonyms MULBUN 'blunt-edged' and MUYMUR 'sour' and JAMAR 'bitter' (1982: 58). This example is of interest for two reasons. Firstly, depending on the direction of transfer (for which no historical data is available), it may constitute an exception to Williams' model. Secondly, it illustrates the reverse association between sweetness and cutting quality found with AMAI in Japanese (cf. 6.4). As Dixon discusses in another context (1982: 59), a plausible explanation for the association here is in terms of (positive) evaluation: just as sweetness is (for the Dyirbal) the most important quality required of a drink, so sharpness is the primary requirement of a cutting implement. This highlights the fact that the mechanism of extension with these terms may be at an affective or evaluative level, as well as at the more specific level of perceived physical qualities (cf. Lehrer, 1978: 119, note 12).

It is also important to consider these extensions from an intralingual semantic perspective (cf. 2.3). Which terms from the basic field are extended, and to what degree are basic intralingual relations preserved? Lehrer (1978) aims to test the hypothesis that, where one or more terms in a semantic system are extended to another domain, the remaining terms are also available for extension to this domain (1978: 96). In general she finds that this is borne out by the behaviour of words in the wine-tasting vocabulary: while gaps certainly occur, the typical pattern is for more than one term, especially closely related terms such as antonyms and synonyms, to appear together in extended domains (1978: 115). Thus, in English SWEET, SOUR and BITTER all appear in the domain of personality, and so on. As far as Japanese is concerned, the evidence as presented in 6.4 is broadly supportive with, however, the notable exception of several areas where AMAI stands as a lone representative. The position of both AMAI and SWEET is noteworthy in several domains. For Japanese, as we have seen in Chapter 5, basic tongue-taste terms are related by contrast, overlain by syntagmatic incompatibility in the case of AMAI. In certain extended uses (within taste, and with reference to strictness of assessment), we find cases where AMAI primarily serves a structural function as the opposite of KARAI, rather than indicating denotational similarity with sweetness. Dirven (1985: 104–5) notes similar extended uses of English SWEET as a 'polarity item' within taste (*sweet water* vs. *salty water, sweet milk* vs. *sour milk*), whence further extensions are made into certain synaesthetic domains (*sweet air*: i.e. 'air free from unpleasant odours'). The significance of these examples is that they clearly illustrate the presence of a purely intralingual mechanism for semantic extension, in addition to extralingual mechanisms based on perceptual or evaluative similarity noted above, and thus highlight the linguistic reality of lexical semantic systems (cf. Lehrer, 1978: 96, 122). In both languages the principal term involved in such cases is 'sweet' (in Japanese, a similar use was also attested for KARAI in the sense of 'dry, i.e. not sweet'), and indeed Dirven's account of the meanings of English SWEET (1985: 104–6) shows close parallels with the representation of the extensions of AMAI in 6.4. In both cases we may distinguish two broad directions, one based on perceptual or affective/evaluative similarity, the other on structural opposition to other basic taste terms; the chief difference is that in Japanese this latter extension is taken further, to the extent of embracing unfavourable qualities in some domains. On this evidence, terms for 'sweet' may be expected to figure prominently in semantic extensions, in part because of their pivotal status as natural opposites in systems of this type, and this topic deserves cross-linguistic exploration. More generally, Lehrer (1978: 115) raises the question whether basic terms are more likely to undergo extension than peripheral items, and this too merits investigation across languages. That historical factors are a potential complication here is suggested by the case of SUPPAI (a relatively recent formation), which shows few, if any, extensions in contrast to English SOUR, and we have noted the relevance of similar factors in the case of OISHII in evaluative taste terms (note 10).

Finally, let us consider briefly the question of semantic extensions into the domain of tongue-taste. Once again, Williams (1976) is a convenient starting-point: he finds that in synaesthetic transfers taste receives terms only from the domain of touch and not, for example, from smell (1976: 463). While we have seen that this may not hold for wider interpretations of 'taste' (6.4), historical evidence for English certainly indicates that touch is a rich source of tongue-taste terms: based on lexicographical citations, BITTER (derived from Old English BĪTAN 'to bite'), DRY, HARSH, HOT, PIQUANT and PUNGENT all emanate from the touch domain (Williams, 1976: 474–6). Swahili -KALI, noted by Dixon as having the senses 'sharp-edged' and 'astringent, bitter' (1982: 58), has perhaps also developed in this way; however, Dixon notes the distinct antonyms BUTU 'blunt-edged' and -TAMU 'sweet', which demonstrates that a similar extension has not occurred in the case of 'sweet'. Indeed, a touch-term origin for 'sweet' seems intuitively less natural than for terms denoting more incisive taste qualities: in the light of our comments above on taste-term systems, one possible source is obviously by development (narrowing) from generalized terms ('good-tasting/sweet'), and Myers suggests just such an origin for 'sweet' in Indo-European (1904: 119). In general Japanese tongue-taste terms in fact show no clear evidence of extension from touch: although authoritative historical lexicographical sources are lacking, basic terms appear to have been historically stable as taste terms for over one thousand years (Nagao, 1982: 42). In 6.4 we have seen a similar contrast in certain extended meanings, in which Japanese employs taste terms (AMAI/KARAKUCHI) whereas English draws upon terms from touch (MILD/DRY) (Field, 1990: 128). While common, extensions from touch are thus by no means ubiquitous and, insofar as origins are known, it is clear that many taste terms are derived from the names of food substances. The following is a sample of such derivations suggested by available sources:

'sweet': milk (Persian: Myers, 1904: 122)
pig-fat (Korafe, Papua New Guinea)[13]
honey (Koko-Yimidir, Australia: Myers, 1904: 123)

'sour': vinegar (Japanese: cf. 6.2)
Brazilian plum (Fijian: Myers, 1904: 123)
tamarind (Toda: Rivers, 1905: 388)

'salty': salt (English)
sea-water (Korafe, Papua New Guinea)[13]
species of dry grass (Hindi: Myers, 1904: 122)

As we have seen, cross-linguistic studies of taste attracted attention at the beginning of the century, but suffered from a lack of sophistication of linguistic theory and method. With the greater insights now available, the time seems ripe for a renewed, more fruitful assault on the language of taste. In this field more than most, progress in our general understanding requires the input of systematic descriptions of everyday language.

Appendix: Summary of Japanese taste terms

This appendix lists in alphabetical order the Japanese taste terms treated in this study. Each term is followed by an English gloss, together with an indication of its main semantic and, where relevant, stylistic properties. Only broad characterizations are provided; fuller details are to be found in the relevant sections of Chapters 4–6.

AKU GA ARU 'has a harshness': AJI II term, applied to e.g. burdock. Synonym of AKUPPOI. –OISHII.

AKUPPOI 'harsh': AJI II term, applied to e.g. burdock. Synonym of AKU GA ARU, stylistically colloquial. –OISHII.

AMAI 'sweet': AJI I term, applied to sugar, etc. +OISHII.

AMAKARAI 'sweet-pungent': AJI I term, applied to dishes cooked with sugar and soy sauce. +OISHII.

AMAMI GA ARU 'has a sweetness': AJI I term, applied to e.g. carrots. +OISHII.

AMAZUPPAI 'sweet-sour': AJI I term, applied to mandarins, strawberries, etc. +OISHII.

AOKUSAI 'grassy-flavoured': AJI III term, applied to e.g. unripe tomatoes. Hyponym of KUSAMI GA ARU 'strong-flavoured', co-hyponym of DOROKUSAI 'muddy-flavoured'. –OISHII.

DOROKUSAI 'muddy-flavoured': AJI III term, applied to carp. Hyponym of KUSAMI GA ARU 'strong-flavoured', co-hyponym of AOKUSAI 'grassy-flavoured'. –OISHII.

HORONIGAI 'pleasantly bitter': AJI I term, applied to e.g. beer. +OISHII.

KAORI GA II 'aromatic': AJI III term, applied to coffee, certain teas, etc. Superordinate of KOOBASHII 'fragrant', contrasts with KUSAMI GA ARU 'strong-flavoured'. +OISHII.

KARAI 'pungent; hot': AJI I term, applied in more general sense ($KARAI_1$) to soy sauce, salt, mustard, etc., in more specific sense ($KARAI_2$) to mustard, etc. $KARAI_1$ is superordinate of SHIOKARAI 'salt'; $KARAI_2$ is superordinate of PIRITTO(-SHITE IRU) 'piquant' and co-hyponym of SHIOKARAI 'salt'. (–)OISHII.

KOOBASHII 'fragrant': AJI III term, applied to dry-roasted or toasted foods. Hyponym of KAORI GA II 'aromatic'. +OISHII.

KUSAMI GA ARU 'strong-flavoured': AJI III term, applied to certain meats, etc. Superordinate of AOKUSAI 'grassy-flavoured' and DOROKUSAI 'muddy-flavoured'. –OISHII.

MAZUI 'bad-tasting': Evaluative term, antonym of UMAI and OISHII 'good-tasting'. Commonly avoided by women with reference to prepared foods.

NIGAI 'bitter': AJI I term, applied to medicinal powders, etc. –OISHII.

NIGAMI GA ARU 'has a bitterness': AJI I term, applied to e.g. saccharine. (–)OISHII.

OISHII 'good-tasting': Evaluative term, antonym of MAZUI 'bad-tasting', synonym of stylistically masculine UMAI.

PIRITTO(-SHITE IRU) 'piquant': AJI I term, applied to e.g. mustard plant. Subsumes two items: adverb PIRITTO 'piquantly' syntagmatically implies $KARAI_2$ 'hot', verb PIRITTO-SHITE IRU 'piquant' is hyponym of $KARAI_2$. +OISHII.

SAMMI GA ARU 'has a sourness': AJI I term, applied to e.g. certain apples. Incompatible with AMAI 'sweet'. +OISHII.

SHIBUI 'astringent': AJI I term, applied to e.g. *shibugaki* (a type of persimmon). –OISHII.

SHIBUMI GA ARU 'has an astringency': AJI I term, applied to e.g. grape skins. (–)OISHII.

SHIOKARAI 'salt': AJI I term, applied to e.g. sea-water. Hyponym of $KARAI_1$ 'pungent', co-hyponym of $KARAI_2$ 'hot'. Contrasts with SHOPPAI 'salty'. –OISHII.

SHOPPAI 'salty': AJI I term, applied to e.g. *kobucha* (a type of seaweed tea). Contrasts with SHIOKARAI 'salt'. (–)OISHII.

SUIMI GA ARU 'has a tartness': AJI I term, applied to e.g. peaches. Compatible only with AMAI 'sweet'. +OISHII.

SUPPAI 'sour': AJI I term, applied to vinegar, lemons, etc. (–)OISHII.

UMAI 'good-tasting': Evaluative term, antonym of MAZUI 'bad-tasting', synonym of OISHII. Stylistically masculine.

Notes

1 Taste and taste terms

1 The following general account is based largely on Amerine at al. (1965), Chapter 2 and Harper (1972), Chapter 6.

2 Bartoshuk (1978: 4–5), McBurney and Gent (1979: 153) and Kuipers (1984: 90) all refer to Aristotle's scheme as containing seven terms: sweet, bitter, sour, salty, astringent, pungent, harsh. However, both *De Anima* (II, 10) and *De Sensu* (IV) have eight-fold schemes; on Aristotle's inclusion of the oily taste, cf. Chapter 7, note 13.

3 Cf. Moskowitz's remarks on these aspects of the eating process:

> Usually, what is contained in the mouth is...a semisolid bolus continually being diluted with and containing saliva. Volatile chemicals emanate from the bolus and are smelled, and the bolus is actively moved about in the mouth, its form changed with each movement, until it is swallowed. (1978: 183)

4 The term 'taste-feel' is from Lehrer, 1978: 98.

5 Here, as often, Myers refers to localities or peoples rather than to specific languages.

6 A question that arises in connection with Maeda's study concerns the semantic effects of the alternations in the case of terms based on 'appropriately salty'. Maeda does not specifically discuss this, although he does point out that the alternations are not fully productive.

7 Cf. the following comment by Gibson on the perceptual role of the mouth in early childhood in contrast with its role in adulthood: 'The human adult does not use his mouth for exploratory haptic touching as the infant seems to do; or, more accurately, he does so only for food substances. "Putting things in the mouth" is tabooed after a certain age' (1966: 143).

8 A good outline of Japanese diet and cuisine can be found in Martin and Martin, 1970.

9 Even in English, highly developed taste vocabularies in specific areas

are of course found in particular sub-cultures, for example amongst wine tasters and connoisseurs (cf. Lehrer, 1975; 1983). A large part of the training of wine tasters can be considered as a course in practical semantics, consisting in acquiring the vocabulary involved and learning to apply it appropriately (in this case, in a way with which other wine tasters agree) to actual instances. Cf. Puisais and Chabanon, 1974: 87, which sets out the content of a well-known wine tasting course.

2 Lexical meaning

1 Lyons uses the term 'sense relations' here: we avoid this term since, in common parlance, different 'senses' of a lexical item may or may not involve differences in intralingual meaning as discussed here. Lehrer (1975: 901) speaks of 'intra-linguistic relationships'.

2 For this aspect of meaning, cf. Lyons, 1968: 443–70 and 1977: 270–317.

3 We draw a typographical distinction throughout this study between lexical items and word-forms (cf. Lyons, 1977: 371; Allan, 1986: 55–8): the former are given in small capitals, the latter in italics.

4 We make a terminological distinction between a lexical semantic system and a lexical semantic field. While paradigmatic relations are criterial for the former, they are not for the latter: 'lexical field' is thus a wider term, taking in, for example, lexical items which occur in response to a common question-frame without necessarily entering into paradigmatic relations. Note that 'field', like 'system', is used here as an intralingual notion, referring to a set of lexical items.

5 Cf. Lyons, 1977: 261–6 and Cruse, 1986: 100–9, 278–82.

6 As Allerton (1984: 26) points out, BLOND also collocates with nouns such as DOLL, STUDENT etc., denoting possessors of hair. In addition, in American English the term has a more technical use in the domain of furniture (Kittay, 1987: 232).

7 There is considerable terminological variation in this area. Our use of the term 'denotation' follows Lyons, 1977: 207. Lyons draws a clear distinction between this relation, which holds between items of the language-system and extralingual categories, and 'reference', which concerns the use of expressions by speakers in particular utterances (cf. 1977: 208).

8 Extralingual meanings, like linguistic meanings in general, are of course internal to the particular language involved. They are also internal in the sense that they involve the world of human experience, rather than simple physical reality.

9 Cf. Lyons' comment on methods of specifying denotation: 'How is the denotation of a lexeme to be specified by the descriptive linguist? The short, practical answer is: in any way that is likely to be successful' (1977: 209).

10 This is a simplified account. For more detailed descriptions in English of Japanese honorifics, cf. Martin, 1975 (Chapters 6 and 22); Harada, 1976; Neustupný, 1978: 213–39.

11 For simplicity, only non-past forms are shown in this and the following table.

12 Neustupný (1978: 235), discussing grammatical aspects of honorifics, speaks of 'hard' and 'soft' forms and is at pains to point out that these items are not to be equated with 'used by males' and 'used by females'. This applies particularly to soft forms, which show a wide spread of usage, ranging from items like OYASAI which are highly marked as used by women (or children) to stylistically neutral terms such as OKANE 'money' (cf. Ide, 1982: 379–80). For the first category we have preferred 'masculine' as more clearly capturing the stylistic flavour of these items. While all are typically used by males, however, our characterization should not be taken to preclude the possibility that some masculine items may be more 'macho' than others; indeed, morphological and collocational evidence for degrees of masculinity will be adduced in Chapter 4.

3 Methodology

1 Examples are commonplace in the literature. Cf., for example, Conklin, 1955, on colour terms; Frake, 1961, on disease terms; and Lyons, 1963, on occupational and epistemological terms.

2 The investigative use of question-and-answer procedures, and the non-trivial nature of asking appropriate questions, are issues that have received attention mainly from ethnoscientists. Cf. Frake, 1964 and Black and Metzger, 1965.

3 Hanunóo in fact appears to be such a language. Cf. Conklin, 1955.

4 Data in Chapter 4 containing masculine variants were not elicited, but constructed and repeatedly checked with the informant.

5 Our use of the unitary label 'isolator' for the particle WA, in place of the usual 'topic'/'contrast' (cf. Kuno, 1973), rests on the view that the basic meaning of WA is to be sought in its opposition to MO ('also, even'); this view is strongly advanced by Martin (1975: 52ff, 224–6), who speaks of 'subduing' (WA) vs. 'highlighting' (MO). We choose the term 'isolator' in contrast to MO's basic function as an 'integrator'.

4 Evaluative taste terms in Japanese

1 By contrast with the present three items, we refer to other taste terms treated in this study (glossed as 'sweet', 'bitter', 'aromatic', etc.) as descriptive taste terms. Even descriptive taste terms, of course, may have an affective value (cf. the discussion in 1.3).

2 In this and similar examples, slashes (*oishii/umai*, etc.) indicate that the forms are interchangeable in the example concerned. The choice between the frames *X wa oishii?* and *X wa umai?* depends on stylistic factors outlined in 2.4.2 and examined in detail below. Responses containing occurrences of UMAI and MAZUI here were not elicited directly from the informant for similar stylistic reasons; cf. Chapter 3, note 4. A further term, BIMI-NA 'good-tasting', is restricted to written Japanese (cf. Nishio, 1972: 103).

3 With eating-places, this frame relates only to the palatability of the food served, and does not seek evaluation of other areas, such as service, atmosphere, etc.: cf. the gloss for (4) below, where the Japanese contains no expression for 'food'.

4 Cf. Backhouse, 1984 for a general survey of Japanese adjectives.

5 For fuller discussions of subjective adjectives cf. Nishio, 1972: 21–42 and Martin, 1975: 359–65. Martin further subdivides the category into 'affective' and 'cathectic' adjectives, according to whether they yield intransitive or transitive -GARU verbs.

6 Martin (1975: 361–5) gives a list of adjectives which produce -GARU verbs, with acceptability ratings based on a survey carried out for some of the terms. He does not include MAZUI, but Nishio (1972: 105) states that all three derivatives occur, and gives an example for MAZUGARU.

7 In relation to smoking articles, let us note that one Japanese equivalent for 'smoke cigarettes' is *tabako o nomu*, NOMU being the normal equivalent to English DRINK.

8 Ide (1982: 379) lists OISHII as a 'beautification honorific' vis-a-vis UMAI, stating that such honorifics serve to beautify speech and in general are most typically used by women. She notes, however, that some items may also be used by men, and that OISHII falls into this category (1982: 380). (Cf. also 2.4.2.)

9 Differences in formality are also reflected in the fact that OISHII, but not UMAI, collocates with honorific verbs and their derivatives: *oishii/ *umai meshiagarikata* 'good-tasting way of consuming (something)'. (MESHIAGARU is the subject-honorific verb corresponding to TABERU 'eat' and NOMU 'drink'; cf. 2.4.2.)

10 It should be noted here that male speakers have available an additional informal option in the phonological variant *umee*. These variants are formed in Tokyo Japanese by the replacement of final *-ai*/*-oi* in adjectival and (negative) verbal forms by *-ee*; they are marked as heavily masculine, and are perhaps to be characterized as 'vulgar' (cf. Wenck, 1966: 18; Ide, 1982: 381).

11 The terms MESHI, KUU and UMAI are brought together and contrasted with their softer stylistic counterparts in Kokuritsu Kokugo Kenkyuujo (1965), a survey of aspects of synonymy based on questionnaires. The relevant section may be translated as follows:

> *Words felt as bad words*
> GOHAN/MESHI TABERU/KUU OISHII/UMAI
> In pairs like those above, while the lefthand members are felt as ordinary (slightly refined) words, the right-hand members are accompanied by a feeling of being (somewhat) bad words, of being rough words. (1965: 25)

Later in the same volume, MESHI and GOHAN are contrasted as follows:

MESHI	GOHAN
(1) Boiled rice, (2) meal; of the three words RAISU, MESHI and GOHAN, MESHI is the oldest term. Gives a somewhat rough impression. Avoided among women and in general in polite speech.	Originally the honorific term for boiled rice, meal. Nowadays in general use as the polite term. Slightly feminine, colloquial. (1965: 51)

12 According to Ishige, Ponapean also generally uses a negative expression of this type (SOHTIOU 'not good-tasting'; cf. IOU 'good-tasting'), although the language also has a distinct word meaning 'bad-tasting' (1983: 18). However, details of usage are not treated in his description.

13 Let us note here that Japanese has no derivational process comparable to UN-prefixation in English for deriving opposites from native adjectives.

14 A nice illustration of sex-associated differences in applicability is found in the weekly magazine *Shuukan Yomiuri* (14 April 1991: 90), where an advertisement for a brand of beer named Malt's (sic) is juxtaposed with an instalment of a cookery column. The beer advertisement features the slogan *Umai mise ni wa Morutsu ga aru* ('In good-tasting establishments, there is Malt's'), with UMAI, while the cookery column is more elegantly entitled *Oishii koyoi* ('Good-tasting (this) evening').

15 It is noticeable that the Japanese equivalents of the English items listed here by Cruse are all subjective adjectives: cf. ATSUI 'hot'/SAMUI 'cold', URESHII 'happy'/KANASHII 'sad', OISHII, UMAI 'nice (of taste)'/MAZUI 'nasty (of taste)', TANOSHII 'pleasant'/IYA-NA 'nasty, unpleasant', TSUMARANAI 'boring, unenjoyable'. Equipollent antonymy amongst opposites in English thus constitutes a semantic reflex of what is a category of much wider linguistic significance in Japanese. Cf. also Chapter 5, note 11.

16 As noted earlier in the section, Nishio (1972) is somewhat ambivalent concerning the subjective status of these terms. While acknowledging the possible existence of examples with a first-person experiencer clearly expressed and the presence of -GARU derivatives, he suggests that the most common function of these terms is nevertheless to denote objective properties of food substances (1972: 104–5). His discussion thus implies that these adjectives belong in an intermediate category, straddling the subjective and objective sub-classes, together with items such as OMOSHIROI and KOWAI which are explicitly discussed in these terms (cf. 1972: 35–6). However, examples such as the following, cited by Nishio in support of an objective analysis, do not appear to pose problems for a subjective interpretation:

Ya, *umai* yo, kono furai. (1972: 104)
'Gee, this tastes good, this fried (fish, etc.).'
Koryaa, *umai* yo. (1972: 104)
'This tastes good'

More relevant may be the general point raised by Nishio that subjective adjectives as a sub-class tend to receive an objective interpretation when used attributively (1972: 34–5). This may also be true of certain adverbial uses: in addition to *oishiku taberu* 'eat in a finding-good-tasting manner', noted above, we find collocations such as *oishiku tsukuru* 'prepare (food) so that it is good-tasting (i.e. good-tastingly)'. These aspects of the behaviour of subjective adjectives in Japanese require further study.

17 According to Hachiya (1983: 33), one meaning of the adjective ISHII (the forerunner of OISHII: cf. 4.5) was also 'skilful'.

18 In English, it is once again noticeable that general evaluators such as GOOD do duty across this whole range (*good Japanese, good at Japanese, go well/badly*, etc.) (cf. 4.1).

5 Descriptive taste terms in Japanese (I)

1 The procedures here thus indicate that (A) is a natural question-frame used in the language to seek the type and level of information that we are interested in. As is clear from other sections of this study, there are of course many other ways of asking questions about taste in Japanese, differing in the area of the taste experience concerned, in generality, etc.

2 Although the question-frame itself is a natural one, it is clear that, insofar as the members of a society do not normally interrogate each other concerning the general taste qualities of basic foodstuffs, questions such as (A1) will not be commonly encountered in everyday conversation. However, they are the natural way of seeking such information in situations where the need arises, as, for example, in conversations with children, in language-teaching, etc.

3 Needless to say, English glosses for Japanese taste terms are to be regarded essentially as labelling devices rather than as exact semantic equivalents.

4 The term 'mimetic' is from Martin; cf. 1975: 1024–5 for some discussion of these items.

5 Cf., as noted in 4.5, the occurrence of UMAMI, derived from the evaluative taste term UMAI, in the phrase UMAMI GA ARU. KARAMI (from KARAI 'pungent') also occurs, but the phrase KARAMI GA ARU is not used by the informant.

6 The expected form here is *sumi*; however, the intrusion of /i/ is found, for example, in *koime* 'on the thick/concentrated side', a comparable derivative from one of the other few monosyllabic adjective bases in the language (*ko-i*). The form *sumi* is in fact listed in many dictionaries but is not used by the informant. By contrast, *suimi* is generally absent from dictionaries; the twenty-volume *Nihon Kokugo Daijiten* lists it as a gloss under *sammi* (Nihon Daijiten Kankookai, 1972–6: vol. 9, 315) and Oono and Hamanishi as a gloss under *sumi* (1985: 230), but in neither does it appear as a separate headword.

7 Cf. the following attested example, in response to a suggestion that a curry was insufficiently hot: *Agichan karagatteru ja nai* 'Aggie's finding it hot, isn't she.'

8 This is clearly a peripheral zone of the taste vocabulary, subject to wide variation across speakers. For Hachiya (1983: 27) EGUI, though historically a taste term, is now chiefly used with extended meanings such as 'cruel, harsh (of disposition), stingy'; however, these are unfamiliar to the informant, and are not generally listed in dictionaries.

9 Since the subjective/objective division applies to adjectives, non-adjectival descriptive taste terms (i.e. -MI GA ARU expressions and AKU GA ARU) are excluded from the discussion here.

10 The acceptability of SUPPAGARU and SHOPPAGARU as formations may be interpreted as one indication of the synchronically simple status of the base adjectives (cf. 5.2).

11 As noted in 4.4, Cruse (1986: 207) classifies English SWEET/SOUR as equipollent antonyms, which are characterized by a subjective dimension in their denotation. Cf. also Chapter 4, note 15.

12 'Contrast set' is also employed in Grandy (1987) and Kittay (1987: 230ff), two recent works in philosophical semantics concerned with semantic fields.

13 *Uun* ('no') co-occurs with simple negation in (B19). It may also occur without the actual negative form being present, i.e. *Uun. Tamanegi wa karai* 'No. Onions are pungent' is also a possible response to (B19). Throughout we regard abbreviated responses of this form as being equivalent to explicit simple negation.

14 Except in baby-talk, where the informant reports that she would use SHOPPAI rather than SHIOKARAI.

15 Tokugawa and Miyajima (1972: 106) suggest that SHOPPAI denotes a mixed taste quality, containing traces of bitterness or sourness as well as saltiness, and suggestions of mixed saltiness and sourness were also made by some respondents (from Nagoya and Osaka) to a written questionnaire on SHOPPAI/SHIOKARAI prepared by the author in 1989. This was distributed during a seminar given in the Department of Japanese Studies at Osaka University; respondents were from various regions of Japan and overall results were accordingly diverse, although several subjects responded in accordance with the informant's system analysed here.

16 Such cases of syntagmatic implication are quite widespread between mimetic adverbs and verbs or adjectives. Cf. the following examples: GARATTO 'totally' and KAWARU 'change', GUSSURI 'soundly' and NERU/NEMURU 'sleep', PERAPERA 'loquaciously, fluently' and SHABERU 'talk', SHIKUSHIKU 'gripingly' and ITAMU 'hurt'/ITAI 'painful', MUZUMUZU 'irritatingly' and KAYUI 'itchy'.

17 Our gloss for SUIMI GA ARU, namely 'has a tartness', is thus to be regarded with particular caution (cf. note 3). We interpret TART as 'pleasantly sour': Lehrer's interpretation is close to this (1978: 98), but dictionaries vary widely in their treatment of the semantics of this item.

18 These two senses were distinguished in the traditional orthography, being represented by distinct characters; however, the character for 'salty' has subsequently dropped out of general use under the post-war simplification of the writing system.

19 The responses to (35) and (36) illustrate the elicited use of MAZUI by the informant. Note that both cases relate to substances which are not normally consumed, as discussed in 4.5.

20 In these examples, the constructions with the conjunctive form (X-*te* Y) have causal meaning 'is X and (thereby) is Y', in contrast with the simple coordinative meaning seen in AND-combinations of AJI I terms. Here, the order descriptive taste term + evaluative taste term is strictly fixed.

21 Cf. the following attested conversational examples with SHOPPAI and NIGAI, which further reflect these characterizations:

Oishii kedo kanari *shoppai* ne.
'They (=potato crisps) taste good but they're pretty salty, aren't they.'

Oishiku nai ne, kono ringo. Sugoku *nigaku* kanjiru.
'These apples don't taste good, do they. They taste very bitter.'

Nigai. Nigai. Mazui.
'Gee, it's bitter. It (=grapefruit) tastes bad.'

6 Descriptive taste terms in Japanese (II)

1 Three of the terms relate morphologically to the adjective KUSAI 'smelly'. We note here that, while KUSAI can be predicated of food-substances (e.g. *Ninniku wa kusai* 'Garlic is smelly'), such a sentence does not constitute an appropriate response to a type-(A) question (*Ninniku wa donna aji ga suru?* 'What kind of a taste does garlic have?') and the term is thus clearly excluded from the field of AJI as defined.

2 The terms SHIOAJI 'salted' and SHOOYUAJI 'soy sauce flavoured', formed by compounding with AJI, are applied to products seasoned with these substances.

3 The main references consulted for this section are Nishio (1972: 98–108, 328–9), Morita (1989: 70–3) and Field (1990, Chapter 5, which also draws upon Backhouse, 1978), together with the dictionaries listed in 5.4.1. Usages were checked with the informant and only those adjudged as being established in everyday use are mentioned here; cf. note 5.

4 As seen below, extensions of this term relate to the domains of saltiness as well as hotness, and we assume that we are dealing with KARAI in its

more general sense (i.e. KARAI$_1$) in this section.

5 Several dictionaries list *shoppai kao o suru* ('make a salty face'), referring to a distressed frowning expression, and Hayashi et al. (1987: 15) mention *suppai kao o suru* ('make a sour face'), referring to a look of frustration. These expressions are not known to the informant, but they suggest a wider availability of (non-sweet) AJI I terms for extension to the domain of negative facial expressions (cf. also *shibui kao o suru* mentioned below). The institutionalization of semantic extensions is often an uncertain area (cf. Lehrer, 1978: 115), in which we may expect variation across speakers. We should also note here the expression *mazui kao o suru* 'make a disapproving face': whether this is more appropriately interpreted as deriving from the basic taste meaning of MAZUI ('the face of someone who has eaten something bad-tasting') or from the sense 'unfavourable, infelicitous' (cf. 4.7) is not clear.

7 Retrospect and prospects

1 Thus in 5.4.1 we noted that the informant rejected the application of AMAKARAI 'sweet-pungent' to tastes involving ginger (as opposed to soy sauce) indicated in Kindaichi et al. (1989: 30), and it is interesting to note that the four compilers of this dictionary are male. A male Tokyo Japanese-speaking acquaintance of the author understands SUIMI GA ARU 'has a tartness' as meaning 'juicy', presumably involving interpretation in terms of the morpheme SUI 'water'.

2 It is possible that this question may decide itself on linguistic grounds: thus, AJI II and AJI III terms will be characterized as non-basic on Berlin and Kay's criteria.

3 Hayashi et al. specifically label SHOPPAI as informal (cf.5.4.2.2), but not SUPPAI (1987: 460, 490).

4 Nishio (1983: 109–11) regards AMAI and KARAI as the most basic Japanese taste terms. His criteria are (a) morphological simplicity, (b) wide geographical distribution, (c) broad semantic range, and (d) wide use in extended meanings. He does not expressly comment on the exclusion of NIGAI or SHIBUI; indeed his discussion omits any mention of NIGAI.

5 Let us recall that weak salt solutions are perceived as having a sweet taste (cf. 1.1).

6 In the light of the 'sour-bitter confusion' found in English (cf. 1.2), the loss of a previously existing distinction should not be excluded as a possible contingency here under particular cultural conditions (such as

reduced exposure to the range of taste qualities concerned).

7 English TART (vis-a-vis SOUR) perhaps belongs here. The unclear status of this term (cf. Chapter 5, note 17) is echoed in Kuipers' analysis of Weyéwa, where PÓLLA 'tart' is listed as a basic term with MAWÍLLURA 'sour' (1984: 92) but elsewhere glossed as 'tart, sour' and described as 'a variety of the *sour* mawíllura sensation' (ibid: 88).

8 SHIBUNIGAI 'astringent-bitter' is attested for an older stage of Japanese, as noted by Nishio (1983: 106–8).

9 English YUMMY, TASTY, PALATABLE, SCRUMPTIOUS, DELICIOUS, etc. are more specific taste evaluators but, as noted in 4.1, these terms are semantically and stylistically restricted in various ways and thus likely to be excluded as basic terms. The presence of YUMMY (YUM) (opposite YUCKY (YUCK)) as a specific taste evaluator in English children's speech is suggestive of the developmental salience of this domain.

10 The Japanese taste evaluators pose an interesting theoretical question with regard to their admissibility as basic terms in view of their stylistic restrictions (cf. 4.5): as we have seen, UMAI in particular is restricted in this way, to male relatively informal speech, and this naturally implies a complementary restriction for OISHII. Such cases present a problem for Berlin and Kay's criteria, which have been further discussed by various scholars. In a study of Russian colour terms, Corbett and Morgan (1988) consider the relevance of derivational potential as a supporting criterion, but conclude that it is unreliable since the number of occurring derivatives of an item will depend partly on the length of time the item has been established in the language, so that historically more recent terms (such as Russian KRASNYJ 'red', and also XOROŠIJ 'good') which otherwise qualify as basic will be excluded. The case of OISHII may be adduced as a further example in support of Corbett and Morgan: as noted in 4.5, it is a relatively recent term, and it is derivationally inhibited vis-a-vis UMAI (4.2), yet its exclusion as a basic term would seem highly counterintuitive. Similar arguments apply to criteria based on occurrence in extended meanings, as employed for example by Nishio (cf. note 4).

11 Similar factors may be at work in the use of evaluative terms more generally: cf., as one example, the presence in English of the lexicalized negative item NO GOOD.

12 Dixon (personal communication, 1991) points out that there are additional dialectal factors involved here, with GULI 'sweet' occurring in northern

dialects of Dyirbal, and GULI 'sharp-edged' in southern dialects; however, the two senses are felt to be related.

13 Information for Korafe is from Cynthia J.M. Farr. Myers (1904: 122–3) mentions associations in other languages between sweet and fatty tastes, and notes that Aristotle places the oily taste close to sweet.

References

Allan, K. 1986. *Linguistic Meaning*. London and New York: Routledge and Kegan Paul

Allerton, D. J. 1984. Three (or four) levels of word cooccurrence restriction. *Lingua*, 63, 17–40

Amerine, M. A., R. M. Pangborn and E. B. Roessler. 1965. *Principles of Sensory Evaluation of Food*. New York, San Francisco and London: Academic Press

Aschmann, H. 1946. Totonac categories of smell. *Tlalocan*, 2, 187–9. (Cited in Sturtevant)

Backhouse, A. E. 1978. Japanese Taste Terms: The Study of a Lexical Field. Unpublished Ph.D. dissertation, University of Edinburgh

1983. The expressive stratum in modern Japanese. *Gengo Kenkyuu*, 83, 61–78

1984. Have all the adjectives gone? *Lingua*, 62, 169–86

Bartoshuk, L. M. 1978. History of taste research. In Carterette and Friedman, eds.

Berlin, B. and P. Kay. 1969. *Basic Color Terms*. Berkeley, Los Angeles and London: University of California Press

Black, M. and D. Metzger. 1965. Ethnographic description and the study of law. *American Anthropologist*, 67, 6, part 2, 141–65

Carterette, E. C. and M. P. Friedman, eds. 1978. *Handbook of Perception Vol. VIA: Tasting and Smelling*. New York, San Francisco and London: Academic Press

Chamberlain, A. F. 1903. Primitive taste-words. *American Journal of Psychology*, 14, 146–53

Conklin, H. C. 1955. Hanunóo color categories. *Southwestern Journal of Anthropology*, 11, 339–44

Corbett, G. and G. Morgan. 1988. Colour terms in Russian: reflections of typological constraints in a single language. *Journal of Linguistics*, 24, 31–64

Coseriu, E. 1971. Die lexematischen Strukturen. In E. Coseriu, *Sprache: Strukturen und Funktionen.* Tübingen: Gunter Narr

Cruse, D. A. 1976. Three classes of antonym in English. *Lingua*, 38, 281–92

1986. *Lexical Semantics.* Cambridge: Cambridge University Press

Dirven, R. 1985. Metaphor as a basic means for extending the lexicon. In W. Paprotté and R. Dirven, eds., *The Ubiquity of Metaphor.* Amsterdam and Philadelphia: John Benjamins

Dixon, R. M. W. 1971. A method of semantic description. In D. D. Steinberg and L. A. Jakobovits, eds., *Semantics.* Cambridge: Cambridge University Press

1982. Where have all the adjectives gone? In R. M. W. Dixon, *Where Have All the Adjectives Gone? and Other Essays in Semantics and Syntax.* Berlin, New York and Amsterdam: Mouton

Field, D. 1990. Patterns of Lexical Synaesthesia in Japanese. Unpublished M. A. dissertation, Australian National University

Frake, C. O. 1961. The diagnosis of disease among the Subanun of Mindanao. *American Anthropologist*, 63, 113–32

1964. Notes on queries in ethnography. *American Anthropologist*, 66, 3, part 2, 132–45

Fujin Kooron. Tokyo: Chuuokooronsha

Fujisawa, S. 1975. Aji no hyoogen no chiikisa. *Gengo Seikatsu*, 286, 52–7

Gibson, J. J. 1966. *The Senses Considered as Perceptual Systems.* Boston: Houghton Mifflin

Goldstein, E. B. 1984. *Sensation and Perception* (2nd edition). Belmont, California: Wadsworth

Grandy, R. E. 1987. In defense of semantic fields. In E. LePore, ed., *New Directions in Semantics.* London: Academic Press

Hachiya, K. 1983. Nihon no mikakugo o megutte. *Bungaku Geijutsu*, 5, 24–40

Hamlyn, D. W. 1968. *Aristotle's De Anima, Books II and III.* Oxford: Clarendon Press

Hanks, P. ed. 1986. *Collins Dictionary of the English Language* (2nd edition). Sydney, Auckland and Glasgow: Collins

Harada, S. I. 1976. Honorifics. In M. Shibatani, ed., *Syntax and Semantics Volume 5: Japanese Generative Grammar.* New York, San Francisco and London: Academic Press

Harper, R. 1972. *Human Senses in Action.* Edinburgh and London: Churchill Livingstone

Harrison, B. 1973. *Form and Content.* Oxford: Blackwell

Hayashi, S., K. Nomoto and F. Minami, eds. 1987. *Reikai Shinkokugo Jiten* (2nd edition). Tokyo: Sanseidoo

Ide, S. 1982. Japanese sociolinguistics: politeness and women's language. *Lingua*, 57, 357–85

Ishige, N. 1978. Harumaheratoo, Galelazoku no shokuseikatsu. *Kokuritsu Minzokugaku Hakubutsukan Kenkyuuhookoku*, 3, 2, 159–271

1983. Mikakuhyoogengo no bunseki. *Gengo Seikatsu*, 382, 14–24

Ishima, T. 1983. Shokuhin ni taisuru hyooka no kisoyooin. In Shibata and Ishige, eds.

Katoo, M. 1966. *Nihongengochizu* kara. *Gengo Seikatsu*, 177, 66–71

Katsumata, S., ed. 1954. *Kenkyusha's New Japanese-English Dictionary* (3rd edition). Tokyo: Kenkyuusha

Kawakami, S. 1987. Taigi to hitei. *Nihongogaku*, 6, 51–9

Kawakita, M. 1960–1. The world of *shibui*. *Japan Quarterly*, 8, 33–42

Kay, P. and C. McDaniel. 1978. The linguistic significance of the meanings of basic color terms. *Language*, 54, 610–46

Kemboo, H., K. Kindaichi, H. Kindaichi and T. Shibata, eds. 1982. *Sanseidoo Kokugo Jiten* (3rd edition). Tokyo: Sanseidoo

Kindaichi, K., T. Shibata, A. Yamada and T. Yamada, eds. 1989. *Shinmeikai Kokugo Jiten* (4th edition). Tokyo: Sanseidoo

Kittay, E. F. 1987. *Metaphor: Its Cognitive Force and Linguistic Structure*. Oxford: Clarendon Press

Kittay, E. and A. Lehrer. 1981. Semantic fields and the structure of metaphor. *Studies in Language*, 5, 31–63

Kokuritsu Kokugo Kenkyuujo. 1965. *Ruigigo no Kenkyuu*. Tokyo: Shuuei Shuppan

Kokuritsu Kokugo Kenkyuujo. 1966a. *Nihongengochizu Kaisetsu: Hoohoo*. (Supplement A to *Nihongengochizu 1*.) Tokyo: Ookurashoo Insatsukyoku.

1966b. *Nihongengochizu Kaisetsu: Kakuzu no Setsumei 1*. (Supplement B to *Nihongengochizu 1*.) Tokyo: Ookurashoo Insatsukyoku

1966–74. *Nihongengochizu 1–6*. Tokyo: Ookurashoo Insatsukyoku

1974. *Nihongengochizu Kaisetsu: Kakuzu no Setsumei 6*. (Supplement to *Nihongengochizu 6*.) Tokyo: Ookurashoo Insatsukyoku

Koono, T. 1974. *Tabemono to Nihonjin*. Tokyo: Koodansha

1977. *Aji to Bunka*. Tokyo: Koodansha

Kuipers, J. 1984. Matters of taste in Weyéwa. *Anthropological Linguistics*, 26, 84–101

Kunihiro, T. 1982. *Imiron no Hoohoo*. Tokyo: Taishuukan

Kuno, S. 1973. *The Structure of the Japanese Language*. Cambridge, Mass. and London: MIT Press

Kyooiku Gijutsu Kenkyuujo, ed. 1986. *Kokugo Kihon'yoorei Jiten.* Tokyo: Kyooikusha

Lehrer, A. 1974. *Semantic Fields and Lexical Structure.* Amsterdam and London: North Holland

1975. Talking about wine. *Language*, 51, 901–23

1978. Structures of the lexicon and transfer of meaning. *Lingua*, 45, 95–123

1983. *Wine and Conversation.* Bloomington: Indiana University Press

Lehrer, A. and K. Lehrer. 1982. Antonymy. *Linguistics and Philosophy*, 5, 483–501

Lindstrom, L. 1986. Kwamera dictionary. *Pacific Linguistics*, C95

Lyons, J. 1963. *Structural Semantics.* Oxford: Blackwell

1968. *Introduction to Theoretical Linguistics.* Cambridge: Cambridge University Press

1977. *Semantics.* Cambridge: Cambridge University Press

Maeda, T. 1978. Choosengo no mikakuhyoogenkeiyooshi. *Choosen Gakuhoo*, 86, 19–35

Mandelbaum, D. G., ed. 1949. *Selected Writings of Edward Sapir in Language, Culture and Personality.* Berkeley and Los Angeles: University of California Press

Martin, S. E. 1975. *A Reference Grammar of Japanese.* New Haven and London: Yale University Press

Martin, P. and J. Martin. 1970. *Japanese Cooking.* Harmondsworth: Penguin Books

Matsubara, M. 1976. Toruko no mura no shokujitaikei. *Kokuritsu Minzokugaku Hakubutsukan Kenkyuuhookoku*, 1, 2, 219–71

McBurney, D.H. and J.F. Gent. 1979. On the nature of taste qualities. *Psychological Bulletin*, 86, 151–67

McIntosh, A. 1961. Patterns and ranges. *Language*, 37, 325–37

Mizutani, O. and N. Mizutani. 1987. *How to be Polite in Japanese.* Tokyo: Japan Times

Morita, Y. 1989. *Kiso Nihongo Jiten.* Tokyo: Kadokawa Shoten

Moskowitz, H. R. 1978. Taste and food technology: acceptability, aesthetics and preference. In Carterette and Friedman, eds.

Myers, C. S. 1904. The taste-names of primitive peoples. *British Journal of Psychology*, 1, 117–26

Nagao, I. 1982. Gomikoo. *Gobun (Nihon Daigaku)*, 55, 34–46

1983. Mikakuyoogo no hensen to bumpu. *Gengo Seikatsu*, 382, 26–33

Nakamura, K., ed. 1965. *Hantaigo Daijiten.* Tokyo: Tookyoodoo

Neustupný, J. V. 1978. *Post-structural Approaches to Language*. Tokyo: University of Tokyo Press

Nida, E. A. 1949. *Morphology* (2nd edition). Ann Arbor: University of Michigan Press

Nihon Daijiten Kankookai, eds. 1972–6. *Nihon Kokugo Daijiten*. Tokyo: Shoogakukan

Nishio, T. 1972. *Keiyooshi no Imi Yoohoo no Kijutsuteki Kenkyuu*. Tokyo: Shuuei Shuppan

1983. Shoku no kankaku o arawasu keiyooshi. In Shibata and Ishige, eds.

O'Mahony, M. and R. Ishii. 1986. A comparison of English and Japanese taste languages: Taste descriptive methodology, codability and the *umami* taste. *British Journal of Psychology*, 77, 161–74

O'Mahony, M. and M. del C. Manzano Alba. 1980. Taste descriptions in Spanish and English. *Chemical Senses*, 5, 47–62

O'Mahony, M. and H. Muhiudeen. 1977. A preliminary study of alternative taste languages using qualitative description of sodium chloride solutions: Malay versus English. *British Journal of Psychology*, 68, 275–8

O'Mahony, M. and T. Tsang. 1980. A preliminary comparison of Cantonese and American-English as taste languages. *British Journal of Psychology*, 71, 221–6

Oono, S. and M. Hamanishi. 1985. *Ruigo Kokugo Jiten*. Tokyo: Kadokawa Shoten

Porzig, W. 1950. *Das Wunder der Sprache*. Berne: Francke

Puisais, J. and R. L. Chabanon. 1974. *Initiation into the Art of Wine Tasting*. Madison, Wisconsin: Interpublish

Rivers, W. H. R. 1905. Observations on the senses of the Todas. *British Journal of Psychology*, 1, 321–96

Robinson, J. O. 1970. The misuse of taste names by untrained observers. *British Journal of Psychology*, 61, 375–8

Rosch, E. 1977. Human categorization. In N. Warren, ed., *Studies in Cross-cultural Psychology Volume 1*. London, New York and San Francisco: Academic Press

Rubin, E. 1936. Taste. *British Journal of Psychology*, 27, 74–85

Sapir, E. 1944. Grading: a study in semantics. *Philosophy of Science*, 11, 93–116. Reprinted in D. G. Mandelbaum, ed.

Scott, G. 1980. Fore dictionary. *Pacific Linguistics*, C62

Seidensticker, E. G. and M. Matsumoto, eds. 1982. *Saishin Nichibei Koogo Jiten* (revised edition). Tokyo: Asahi Shuppansha

Shibata, T. 1957. 'O' no tsuku go, tsukanai go. *Gengo Seikatsu*, 70, 40–9

Shibata, T. and N. Ishige, eds. 1983. *Shoku no Kotoba*. Tokyo: Domesu Shuppan

Shuukan Yomiuri. Tokyo: Yomiuri Shimbunsha

Slobin, D. I. 1971. *Psycholinguistics*. Glenview, Illinois: Scott, Foresman

Sturtevant, W. C. 1964. Studies in ethnoscience. *American Anthropologist*, 66, 3, part 2, 99–131

Tokugawa, M., ed. 1979. *Nihon no Hoogenchizu*. Tokyo: Chuuookooronsha

Tokugawa, M. and T. Miyajima, eds. 1972. *Ruigigo Jiten*. Tokyo: Tookyoodoo

Wenck, G. 1966. *The Phonemics of Japanese: Questions and Attempts*. Wiesbaden: Otto Harrassowitz

Westney, P. 1986. Notes on scales. *Lingua*, 69, 333–54

Wierzbicka, A. 1990. The meaning of color terms: semantics, culture, and cognition. *Cognitive Linguistics*, 1, 99–150

Williams, J. M. 1976. Synaesthetic adjectives: a possible law of semantic change. *Language*, 52, 461–78

Wyburn, G. M., R. W. Pickford and R. J. Hirst. 1964. *Human Senses and Perception*. Edinburgh and London: Oliver and Boyd

Yamada, S. 1972. Gendai Miyakohoogen mikakugoikoo. *Okinawa Bunka*, 39, 30–7

Index

ABURAKKOI, 32
acquired tastes, 3, 4, 10, 14, 17
adjectives, 15, 169, 178
  evidential, 40, 41, 53, 69, 95, 155, 157
    see also -SOO-NA
  i-adjectives, 39, 81, 96
  subjective, 39, 50, 65–9, 82, 94–5, 162, 178, 180, 182
AJI, 63, 73, 74, 79, 84, 85, 87, 114, 148, 149, 150, 155, 158, 159, 160, 161, 164, 183
AKU GA ARU, 77, 80, 82, 83, 141, 150, 158, 182
AKUPPOI, 77, 80, 81, 82, 83, 86, 87, 141, 150, 158
Algonquian languages, 6
Allan, K., 28, 176
Allerton, D. J., 22, 176
AMAI, 16, 32, 75, 77, 78, 80, 81, 84, 85, 86, 87, 89, 91, 92, 93, 94, 95, 96, 97, 98, 99, 100, 101, 102, 108, 110, 111, 112, 113, 115, 116, 117, 118, 119, 120, 121, 128, 129, 130, 131, 132, 133, 134, 135, 136, 137, 138, 139, 140, 151, 152, 153, 154, 155, 156, 157, 158, 162, 165, 169, 170, 171, 172, 184
AMAKARAI, 75, 80, 81, 87, 89, 91, 93, 94, 95, 101, 102, 108, 120, 121, 128, 129, 130, 131, 132, 133, 134, 136, 137, 138, 139, 140, 165, 166, 184
AMAMI GA ARU, 79, 80, 82, 87, 88, 90, 108, 109, 111, 114, 128, 129, 130, 132, 133, 136, 139, 140
AMATTARUI, 83
AMAZUPPAI, 16, 17, 75, 80, 81, 84, 85, 86, 87, 89, 91, 92, 95, 96, 97, 98, 100, 101, 102, 108, 116, 117, 118, 120, 121, 128, 129, 130, 131, 132, 133, 134, 136, 137, 138, 139, 140, 157, 158, 165, 166
Amerine, M. A., 5, 175
Anatolian Turkish, 10, 167
antonymy, 15, 20, 41, 42–51, 59, 60, 69, 72, 97, 99, 134, 152, 153, 154, 156, 168, 170, 171, 180, 182
AOKUSAI, 78, 81, 82, 87, 113, 142, 143, 144, 145, 146, 147, 148, 157
Aristotle, 4, 175, 186
Aschmann, H., 9, 25
astringency, 4, 13, 94, 141, 143, 150, 160, 166
ASTRINGENT, 98

Backhouse, A. E., 81, 178, 183
Bartoshuk, L. M., 2, 175
Berlin, B., 12, 24, 25, 161, 164, 165, 184, 185
BETABETA-SHITE IRU, 63
BIMI-NA, 178
BITTER, 11, 14, 15, 98, 134, 169, 170, 171
BITTERSWEET, 158
bitter taste, 2, 3, 137
Black, M., 177
BOSOBOSO-SHITE IRU, 64

Cantonese, 11
Chabanon, R. L., 176
Chamberlain, A. F., 6
chemical senses, 1, 2
collocation, 21, 22, 23, 28, 40–1, 45, 46, 47, 55, 57, 61, 72, 105, 116, 153, 154, 155, 157, 158, 159, 163, 169
colour, 6, 9, 12–14
colour terms, 6, 7, 12, 14, 15, 23, 24, 25, 26, 33–4, 80, 97, 98, 99, 112, 115, 126, 167, 185

Conklin, H. C., 177
contrast, 98–9, 100, 103, 108, 112, 116, 126, 138, 139, 147, 151, 153, 170, 182
see also incompatibility
Corbett, G., 185
Coseriu, E., 21
Cruse, D. A., 20, 28, 29, 38, 44, 49, 50, 51, 69, 176, 180, 182
DELICIOUS, 14, 15, 38, 41, 72, 185
denotation, 5, 23, 24, 25, 26, 28, 29, 33, 34, 44, 47, 50, 94, 126, 137, 139, 153, 156, 166, 170, 176, 177, 182
denotational axioms, 16–17, 25, 26, 35, 88, 130, 131, 161
diet, 13, 17, 74, 91, 93, 104
Japanese, 17–18, 70, 118, 175
Dirven, R., 169, 170, 171
Dixon, R. M. W., 15, 28, 39, 169, 170, 171, 185
DOROKUSAI, 78, 81, 82, 87, 142, 143, 144, 145, 146, 147, 148, 158
DRY, 171, 172
Dyirbal, 170, 186

EGARAPPOI, 83
EGOI, 83
EGUI, 83, 141, 181
English, 5, 11, 13, 15, 16, 19–20, 38, 134, 151, 157, 168, 171, 172, 185
extended meanings, *see* semantic extension

Farr, Cynthia J. M., 186
Field, D., 151, 172, 183
Fijian, 172
FLAVOUR, 164
focal exemplars, 16, 23, 24, 25, 26, 89–91, 93, 123, 141, 161
Fore, 167, 168
Frake, C. O., 98, 177
Fujin Kooron, 35, 45, 46, 47, 48, 52, 53, 54, 55, 56, 57
Fujisawa, S., 9, 104

Galelan, 10, 134, 167
-GARU, 39, 40, 41, 66–7, 82, 94–5, 178, 180
Gent, J. F., 2, 175
Gibson, J. J., 4, 5, 175
Goldstein, E. B., 3
GOOD, 14, 38, 168
Grandy, R. E., 182
gustation, 4, 5, 12, 13, 64, 73, 94, 143, 150, 160, 164, 165, 166

Hachiya, K., 9, 57, 72, 81, 180, 181
Hamanishi, M., 181
Hamlyn, D. W., 4
Hanks, P., 14
Harada, S. I., 177
Harper, R., 2, 5, 11, 175
Harrison, B., 26
HARSH, 171
harshness, 143, 150, 160
Hayashi, S., 92, 104, 105, 123, 157, 184
Hindi, 172
honorifics, 29, 177, 178, 179
HORONIGAI, 76, 80, 81, 82, 87, 90, 91, 93, 95, 101, 102, 108, 120, 121, 128, 129, 130, 131, 132, 133, 134, 135, 136, 137, 138, 139, 140, 158, 162, 165, 166
HOT, 152, 157, 171
hyponymy, 15, 20, 28, 29, 99, 103, 107, 108, 122, 125, 126, 130, 138, 144, 145, 147, 165

Ide, S., 55, 177, 178, 179
IGARAPPOI, 83
II, 38, 41, 49, 50, 69, 70, 168
implication, 19, 42, 43, 44, 49, 71, 116
syntagmatic, 21, 22, 105, 182
incompatibility
paradigmatic, 15, 20, 23, 28, 29, 97–8
see also contrast
syntagmatic, 22, 41, 99, 130, 132, 133, 139, 140, 147, 148, 151, 162, 170
Ishige, N., 10, 35, 45, 46, 48, 58, 134, 151, 167, 179
Ishii, R., 11, 61, 62
Ishima, T., 47, 48, 61

Japanese dialects, 9, 57, 82, 104, 123

KAORI, 82, 149, 150
KAORI GA II, 16, 63, 64, 78, 81, 85, 86, 87,

142, 143, 144, 145, 146, 147, 148
KARAI, 16, 75, 76, 77, 80, 81, 83, 84, 85, 87, 89, 91, 92, 93, 94, 101, 102, 103, 104, 105, 106, 107, 108, 113, 114, 120, 121, 122, 123, 124, 125, 126, 127, 128, 130, 131, 132, 133, 134, 135, 136, 137, 138, 139, 140, 151, 152, 153, 154, 156, 162, 165, 169, 170, 171, 183, 184
KARAMI, 181
Katoo, M., 9, 72
Katsumata, S., 155
Kawakami, S., 96, 97
Kawakita, M., 159
Kay, P., 12, 24, 25, 26, 112, 161, 164, 165, 184, 185
Kemboo, H., 92, 93
Kindaichi, K., 92, 93, 94, 141, 184
Kittay, E., 21, 176, 182
Koko-Yimidir, 172
Kokuritsu Kokugo Kenkyuujo, 9, 57, 82, 91, 92, 104, 123, 179
KOOBASHII, 32, 78, 81, 82, 87, 142, 143, 144, 146, 147, 148
Koono, T., 35, 48, 52, 53, 58, 149
Korafe, 172, 186
Korean, 10, 134
KORIKORI-SHITE IRU, 32, 163, 164
Kuipers, J., 9, 80, 163, 167, 170, 175, 185
Kunihiro, T., 9
Kuno, S., 67, 177
KUSAI, 81, 82, 183
KUSAMI GA ARU, 78, 80, 82, 84, 85, 87, 142, 143, 144, 145, 146, 147, 148
Kwamera, 167, 168
Kyooiku Gijutsu Kenkyuujo, 52, 56

Lehrer, A., 8, 9, 13, 14, 20, 21, 27, 65, 70, 99, 156, 158, 169, 170, 171, 175, 176, 182, 184
Lehrer, K., 99
lexical field, 19, 33, 34, 35, 73, 82, 85, 148, 163, 176, 182
lexical system, 12, 20, 23, 26, 27, 32, 33, 34, 73, 85–8, 171, 176
Lindstrom, L., 167
Linguistic Atlas of Japan, 9, 34, 57, 91, 104, 123
Linnaeus, 4
Lyons, J., 19, 20, 21, 41, 42, 43, 44, 97, 98, 176, 177

Maeda, T., 10, 134, 175
Malay, 11, 12
Mandelbaum, D. G., 43
Manzano Alba, M. del C., 11
Martin, J., 175
Martin, P., 175
Martin, S. E., 40, 82, 94, 177, 178, 181
Matsubara, M., 10
Matsumoto, M., 159
MAZUI, 37, 38, 39, 40, 41, 45, 46, 47, 48, 49, 50, 51, 57, 58, 59, 60, 61, 62, 63, 64, 65, 67, 69, 70, 71, 72, 97, 134, 135, 136, 164, 168, 178, 183, 184
McBurney, D. H., 2, 175
McDaniel, C.,12, 26, 112
McIntosh, A., 21
meaning
  extralingual, 22–6, 62–71, 85–95, 103, 116, 117, 118, 119, 120, 123–4, 126, 130, 131, 138, 141, 142–3, 153, 154, 156, 157, 162, 171, 176
    see also denotation, denotational axioms, focal exemplars, taste norms
  intralingual, 19–22, 23, 26, 27, 28, 29, 34, 40–51, 95–140, 141–2, 144–8, 156, 170, 171, 176
    see also lexical system, semantic relations
  and style, 27–9
  see also semantic extension
metallic taste, 2
Metzger, D., 177
-MI, 40, 61, 82
MILD, 153, 172
Miyajima, T., 62, 71, 182
MIZUPPOI, 32
Mizutani, N., 55
Mizutani, O., 55
monosodium glutamate, 11, 62
Morgan, G., 185
Morita, Y., 62, 71, 151, 157, 183
Moskowitz, H. R., 3, 175
Muhiudeen, H., 11

Myers, C. S., 6, 7, 8, 94, 166, 167, 172, 175, 186

Nagao, I., 9, 62, 172
Nakamura, K., 151
NAMANURUI, 64
negation, 95–7, 98–9
Neustupný, J. V. 177
NICE, 38, 168
Nida, E. A., 9
NIGAI, 76, 77, 80, 81, 83, 87, 90, 91, 92, 93, 94, 95, 96, 97, 98, 99, 101, 102, 108, 109, 111, 113, 114, 115, 120, 121, 128, 129, 130, 131, 132, 133, 136, 137, 138, 139, 140, 151, 158, 162, 165, 183, 184
NIGAMI GA ARU, 77, 80, 82, 87, 88, 90, 108, 109, 111, 113, 119, 128, 129, 130, 132, 133, 136, 139, 140, 149
Nihon Daijiten Kankookai, 181
Nishio, T., 9, 57, 62, 66, 67, 81, 83, 91, 92, 94, 153, 157, 178, 180, 183, 184, 185

Öhrwall, H., 2
OISHII, 31, 37, 38, 39, 40, 41, 42, 45, 46, 47, 48, 49, 50, 51, 52, 53, 54, 55, 56, 57, 58, 59, 60, 61, 62, 63, 64, 65, 66, 67, 68, 69, 70, 71, 72, 95, 118, 134, 135, 136, 163, 164, 168, 171, 178, 179, 180, 183, 185
olfaction, 2, 4, 5, 12, 13, 64, 143, 150, 160, 164
  see also smell
O'Mahony, M., 11, 61, 62
OOAJI-NA, 61
Oono, S., 181
Ozu, Y., 52

pain, 4, 64
PALATABLE, 38, 185
PARIPARI-SHITE IRU, 64
Persian, 172
PIQUANT, 157, 171
PIRITTO(-SHITE IRU), 63, 64, 76, 80, 82, 87, 90, 101, 105–7, 108, 120, 121, 122, 123, 124, 125, 126, 127, 128, 130, 132, 133, 136, 137, 138, 139, 140, 166
Ponapean, 10, 151, 167, 168, 179
Porzig, W., 21
primary tastes, 2, 160
proximate senses, 13, 14, 161
Puisais, J., 176
pungency, 13, 94, 143, 150, 156, 157, 160, 166
PUNGENT, 157, 171

Rivers, W. H. R., 8, 166, 172
Robinson, J. O., 10, 11
romanization, 35–6
Rosch, E., 24
Rubin, E., 4, 5

-SA, 40, 61, 82
SALTY, 15, 157, 166
salty taste, 2, 3
SAMMI GA ARU, 78, 80, 82, 86, 88, 90, 109, 110, 111, 113, 118–19, 128, 129, 130, 132, 133, 134, 136, 137, 139, 140
Sanskrit, 8, 94
Sapir, E., 43
SAPPARI-SHITE IRU, 32
scalarity, 99, 103
Scott, G., 167
SCRUMPTIOUS, 38, 185
Seidensticker, E. G., 158
Seligmann, C. G., 6
semantic extension, 8, 9, 13, 15, 26–7, 71–2, 83, 151–9, 163, 169–72, 184
semantic field, *see* lexical field
semantic relations
  paradigmatic, 19–21, 27, 28, 41–51, 71, 144–7, 176
    see also antonymy, contrast, hyponymy, incompatibility, synonymy
  syntagmatic, 14, 21–2, 28, 40–1, 119–22, 127–9, 134–7, 147–8
    see also collocation, implication, incompatibility
semantic system, *see* lexical system
Shibata, T., 31, 35, 45, 46, 48, 58
SHIBUI, 16, 32, 76, 80, 81, 83, 87, 90, 91, 92, 93, 94, 96, 97, 98, 99, 108, 109, 111, 115, 121, 128, 129, 132, 133, 136, 138, 139, 140, 141, 151, 158, 165, 166, 184
SHIBUMI GA ARU, 77, 80, 82, 88, 90, 108,

109, 111, 113, 114, 115, 128, 129, 132, 133, 136, 139, 140
SHIBUNIGAI, 185
SHIKOSHIKO-SHITE IRU, 163, 164
SHIOAJI, 183
SHIOKARAI, 76, 80, 81, 87, 89, 91, 93, 95, 99, 101, 102, 103, 104, 105, 107, 108, 120, 121, 122, 123, 124, 125, 126, 127, 128, 130, 131, 132, 133, 136, 137, 138, 139, 140, 165, 182
SHITAZAWARI GA II, 16, 18
SHOOYUAJI, 183
SHOPPAI, 76, 80, 81, 82, 87, 89, 91, 93, 95, 99, 101, 103, 104, 105, 108, 120, 121, 122, 125, 126, 127, 128, 130, 131, 132, 133, 135, 136, 137, 138, 139, 140, 165, 166, 182, 183, 184
Shuukan Yomiuri, 179
skin senses, 2
Slobin, D. I., 13
smell, 1, 2, 4, 161, 169, 171
  see also olfaction
smell terms, 9, 158
-SOO-NA, 40, 61, 95
SOUR, 11, 15, 20, 23, 44, 98, 134, 151, 170, 171, 182, 185
sour taste, 2, 3
Spanish, 11, 12
Sturtevant, W. C., 6, 9, 12, 13, 25
style, 27–31, 34, 35, 41, 42, 51–62, 71, 72, 81, 82, 104, 141
  extra-formal, 30, 54, 61
  masculine, 31, 34, 51, 53, 54, 55, 56, 58, 60, 62, 177, 179
SUI, 82, 92, 109, 158, 165
SUIMI GA ARU, 37, 38, 63, 64, 77, 80, 82, 86, 88, 90, 91, 109, 110, 112, 113, 115–18, 119, 128, 129, 130, 131, 132, 133, 134, 136, 137, 139, 140, 182, 184
SUJIPPOI, 32
SUKI-NA, 69, 70
SUKKAI, 82
SUMI, 181
SUPPAI, 16, 17, 71, 75, 80, 81, 82, 86, 87, 89, 91, 92, 93, 94, 95, 98, 99, 100, 101, 102, 108, 109, 110, 111, 115, 116, 117, 121, 128, 129, 130, 131, 132, 133, 134, 135, 136, 137, 138, 139, 140, 151, 157, 165, 171, 182, 184
SUPPAMI, 82
Swahili, 171
SWEET, 5, 14, 15, 16, 19, 20, 21, 22, 23, 26, 44, 98, 134, 151, 156, 169, 170, 171, 182
sweet taste, 2, 3, 10, 137
synonymy, 20, 28, 29, 41–2, 62, 107, 141, 170, 179

TART, 182, 185
TASTE, 5, 20, 21, 22, 38, 159, 164
  affective aspects of, 3, 14, 137, 162, 163, 166
  biological function of, 3, 10, 14
  vs. colour, 6, 9, 12–14
  cultural aspects of, 3, 10, 13, 17, 65, 70–1, 157, 161–2, 163
  individual variation in, 3, 14, 34, 71, 137, 162
  scientific study of, 1–5
  sensitivity to, 2, 3
  see also acquired tastes, gustation, primary tastes, tongue-taste
taste blindness, 3
taste buds, 1, 2
taste cells, 1, 4
taste combinations, 111, 112–13, 166
  see also taste mixtures
taste expectations, *see* taste norms
taste-feel, 5, 163, 164, 175
taste mixtures, 2, 3
  see also taste combinations
taste norms, 14, 16–17, 35, 70–1, 88–91, 116, 137, 161
taste scales, 3
taste terms
  basic, 164–8, 170, 172, 185
  vs. colour terms, 12–15, 39–40, 66, 68, 69, 80, 94, 98, 99, 112–13, 160–2
  delimitation of, 5, 32–4, 37–8, 73–9, 163, 164, 165, 166, 169
  descriptive, 15, 17, 38, 39, 40, 65, 72, 73–159, 162, 163, 167, 168, 169, 178
  in discourse, 9, 163

evaluative, 15, 34, 37–72, 94, 95, 134, 140, 147, 157, 162, 163, 164, 167–8, 169
poverty of, 13, 161
see also references to individual languages
TASTY, 38, 185
Telugu, 15
Toda, 8, 166, 172
Tokugawa, M., 62, 71, 81, 182
tongue-taste, 150, 157, 160, 161, 164, 165, 166, 167, 168, 169, 170, 171, 172
Torres Strait languages, 6
Tsang, T., 11

UMAI, 31, 32, 37, 38, 39, 40, 41, 42, 45, 46, 47, 48, 49, 50, 51, 52, 53, 54, 55, 56, 57, 58, 59, 60, 61, 62, 63, 64, 65, 69, 70, 71, 72, 97, 134, 136, 164, 168, 178, 179, 180, 181, 185
UMAMI, 40, 41, 61, 62, 181
USUAMAI, 83

Wenck, G., 179
Westney, P., 99
Weyéwa, 9, 163, 167, 169, 185
Wierzbicka, A., 12
Williams, J. M., 26, 151, 154, 157, 158, 169, 170, 171
Wyburn, G. M., 3

Yamada, S., 9, 83, 104
YUCKY, 185
YUMMY, 38, 185

For EU product safety concerns, contact us at Calle de José Abascal, 56–1°, 28003 Madrid, Spain or eugpsr@cambridge.org.

www.ingramcontent.com/pod-product-compliance
Lightning Source LLC
LaVergne TN
LVHW090359210926
848480LV00002B/390

* 9 7 8 0 5 2 1 0 2 3 2 2 1 *